ENVIRONMENTAL RESEARCH ADVANCES

ENVIRONMENT, CLIMATE CHANGE AND GREEN ENTREPRENEURSHIP

A JOURNEY TOWARDS SUSTAINABLE DEVELOPMENT

ENVIRONMENTAL RESEARCH ADVANCES

Additional books and e-books in this series can be found on Nova's website under the Series tab.

ENVIRONMENTAL RESEARCH ADVANCES

ENVIRONMENT, CLIMATE CHANGE AND GREEN ENTREPRENEURSHIP

A JOURNEY TOWARDS SUSTAINABLE DEVELOPMENT

FARAHA NAWAZ

nova science publishers
New York

Library of Congress Cataloging-in-Publication Data

ISBN: 978-1-53619-186-8

Published by Nova Science Publishers, Inc. † New York

CONTENTS

PREFACE

This book aims to analyse the nexus between climate change and green entrepreneurship. This book argues that green entrepreneurship is one of the key strategies for mitigating the adverse impacts of climate change. However, adopting green entrepreneurship for mitigating climate change is a challenging journey towards sustainable development. This book also highlights the prospects and challenges of green entrepreneurship in Bangladesh in the context of environment and climate change while investigating how the ideas of green entrepreneurship have been perceived and achieved. For this study, the author has selected those green entrepreneurs whose main target is both recycling waste products and the use of renewable energy sources. Green entrepreneurship has emerged as a symbiotic relationship between the environment and entrepreneurship. Other main targets of green entrepreneurship include recycling waste products and increasing the use of renewable energy sources to mitigate climate change. The core idea of green entrepreneurship is to provide positive environmental outcomes when producing goods and services. One of the most important aims of green entrepreneurship is to protect the natural environment from all adverse effects and diminish the negative consequences of climate change. This book is based on both primary and secondary data. Primary data has been collected from a number of green entrepreneurs, non-green entrepreneurs, government officials and Key

Informant Interviewees (KII). Secondary data has been collected from various publications, journals, books, articles, and websites. The findings of this book suggest that there are many challenges facing the expansion of green entrepreneurship in Bangladesh including financial, behavioral, and organizational challenges.

ACKNOWLEDGMENTS

My book on Environment, climate change and green entrepreneurship: a journey towards sustainable development was created to meet an academic purpose. Writing a book is, of course, a very challenging journey. In accomplishment of this book, many people have bestowed upon me their blessings and their wholehearted support. It is my moral obligation to express my sincere gratitude to the people who have assisted me along the way. Firstly, my heartfelt gratitude goes to Dr. A. K. M Ahsan Ullah from the University of Brunei Darussalam for his support that has been so instrumental in enabling me to reach my desired goal. He also generously provided me with recent reading materials that have supported and enriched my book. I am further privileged to have a number of international professors' cordial academic support that has been a tremendous help as I worked through all the critical issues, to complete my work from the very beginning to the end. Without their cooperation I would have been unable to write this book to its current standard. I am also very much indebted to my international colleague Diana Fox for her outstanding cooperation, encouragement and endorsement for this book.

I am also thankful to my friend Mohammod Tauheed and Sarah- Jane who have assisted me in the acquisition of innovative techniques that contributed to the quality of my book. I am also obliged to all the participants in my research: the green entrepreneurs, non- green entrepreneurs, bank officials, Key Informant Interviewees (KIIs) and government officials. I am grateful to everyone who has sacrificed their valuable time for me during my field study. I am indebted to my research assistants and data collectors who assisted me to conduct my field work. Without their co-operation my field work journey would have been very challenging. I am truly grateful to beloved parents and my husband for their constant inspiration, encouragement, suggestions, and emotional support which are all very important for producing a high-quality book. A special thanks to my little daughter Fiona whose smile gave me the courage and confidence to complete this challenging journey. Above all, I offer the flowers of gratitude through prayer to the Almighty 'ALLAH' for paving each stone on my way to success. Finally, let me conclude by articulating that I have tried my very best to produce a precise, resourceful, exceptional and informative book and I accept full responsibility for any mistakes that I may have inadvertently made.

ACRONYMS

BCCTF	Bangladesh climate change trust fund
BCCRF	Bangladesh Climate Change Resilience Fund
BCCSAP	Bangladesh Climate Change Strategy Action Plan
CBOs	Community Based Organizations
DFID	Department for International Development
DOE	Department of Environment
ECOSOC	Economic and Social Council
FGDs	Focus Group Discussions
GDP	Gross Domestic Product
GPRB	Government of the People's Republic of Bangladesh
IDCOL	Infrastructural Development Company Limited
ILO	The International Labor Organization
IPCC	Intergovernmental Panel on Climate Change
IUCN	International Union for Conservation of Nature
MDGs	Millennium Development Goals
MoEF	Ministry of Environment and Forest
MOP	Ministry of Planning
NCS	National Conservation Strategy
NEMAP	National Environment Management Action Plan
NAMA	Nationally Appropriate Mitigation Action
NGOs	Non-Government Organizations

SBS	Sustainability Balanced Scorecard
SERDA	Sustainable and Renewable Energy Development Authority
SME	Small Medium Enterprise
SDGs	Sustainable Development Goals
UN	United Nations
UNFCCC	United Nations Framework Convention on climate change
UNEP	United Nations Environmental Program
UNCED	United Nations Conference on Environment and Development
UNEP	United Nations Environment Program
WB	World Bank
WCED	World Commission on Environment and Development

Chapter 1

PROLOGUE

Bangladesh has made strong economic gains since its independence in 1971. Over the past thirty years growth has accelerated, with growth increasing by 6% per year on an average over the past eight years. Bangladesh aspires to be a middle-income country by 2020 and is capable of controlling inflation with some exceptions including high inflation during natural disasters and global price increases. Helal and Hossain (2013) reported that public expenditure in Bangladesh is also the lowest in the world. Bangladesh has made enormous socio-economic progress since its independence. The country has achieved most of its Millennium Development Goals (MDG) targets particularly in poverty reduction, child mortality, maternal mortality, and gender equality in education (GPRB, 2008 cited in Sachs, 2012).

Despite all this positive development and economic growth, Bangladesh is one of the most vulnerable countries in the world to the impacts of climate change. Bizikova (2007) stated that climate change refers to a substantial change in temperature, precipitation, or wind lasting for a long period of time, usually several years. Therefore, the core idea of climate change is an increase or decrease in temperature as a result of greenhouse gas emissions into atmosphere. The consequences of climate change are an increase in global air and ocean temperature on average, widespread melting of snow and rising global average sea level.

The Bay of Bengal's 'funnel shape' combined with its connection with the rivers of Bangladesh creates a constant hazardous situation for five coastal areas of the country. The high frequency of natural disasters also hinder the country's ability to achieve its desired rates of economic growth, evidenced particularly by cyclones Gorky in 1991, Sidr in 2007 and Aila in 2009. DFID (2011) reported that climate change is expected to cause increasingly frequent and severe floods, tropical cyclones, storm surges, and droughts. Similarly, The Asia Foundation (2012) also reported that Bangladesh is one of the most vulnerable countries in the world regarding climate change due to its high population density, low income, weak infrastructure and its geographical position on a low-lying delta.

Beg et al. (2011) argue that in developing countries climate change is often not taken into serious consideration by environmental or financial policy agendas but affects the prospect for the development of those countries. Developing countries, however, are beset with huge challenges including overpopulation, illiteracy, unemployment, and malnutrition while also being the most vulnerable to the negative impacts of climate change. Beg et al. (2011) further argue that it is essential to recognize how climate change is likely to influence other development areas and that recognizing climate change as a crucial factor is the first step toward building cost effective strategies and integrated, institutional capacity in developing countries.

Climate Change issues have traditionally broken into two basic categories: climate change mitigation and climate change adaptation. Climate change mitigation focuses on the causes of climate change and it refers to the reduction of greenhouse gas emissions into the atmosphere that are heating our planet. Mitigation strategies include adopting renewable energy sources such as solar, wind, hydro, biofuels, and the sustainable use of land and forests. On the other hand, climate change adaptation focuses on the effects of climate change and how best to manage the consequences of climate change. However, more integrated approaches are needed. This book argues that green entrepreneurship has been regarded as one of the key strategies to mitigate climate change. The following figure highlights the health impact of climate change.

The health impact of climate change is severe. The figure depicts some weather extremes such as heatwaves, winter cold has more direct effect one health. There are some other extreme weather events such as floods, cyclones, droughts and increased production of air pollutant such as spores and moulds also have direct negative effect on health. For dealing with the adverse effect of climate change, some mitigation and adaptation actions can be taken. Bhandari et al., Golkany, 2007 cited in Beg et al. (2011) argued that in order to promote integrated adaptation and mitigation measures, the emphasis should be given to sustainable development. They also stated that a combination of climate change mitigation and adaptation depends on the extent and rate of climate change in a particular local context.

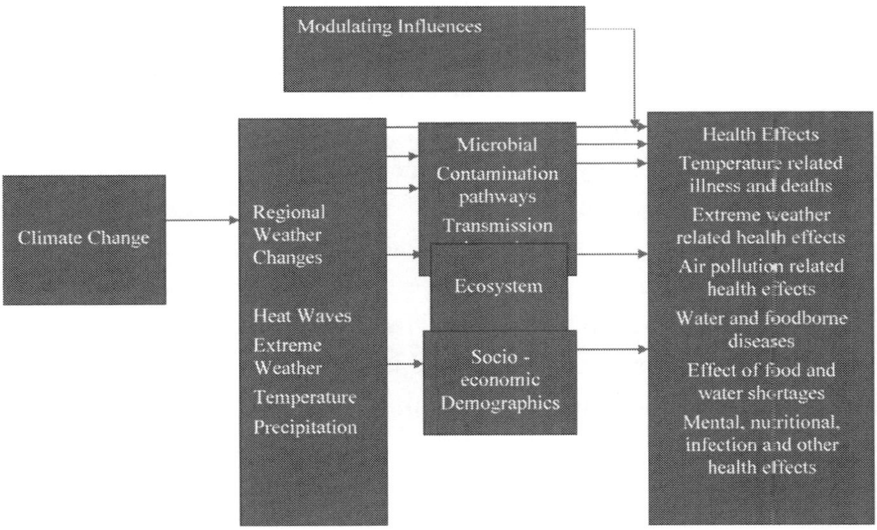

Source: Adapted from Patz et al., 2000, Adedej et al. (2014:119).

Figure 1. Health impact of climate change.

IPCC (2007b) published Working Group II 'Summary for Policymakers' where they identified the linkages between climate change and sustainable development. The IPCC Fourth Assessment has identified adaptation and mitigation as concurrent themes across the Working Group II and III reports. Wilbanks and Sathaye 2007, citing Beg et al. (2011) found that an inclusive model of climate change mitigation and adaptation that enlisted a number of potential sectors such as forestry, agriculture, water and insurance was essential to reduce climate change's adverse effects. By emphasizing local context and local capacities, it would be easier to offer realistic mitigation and adaptation pathways to countering climate change. This book argues that green entrepreneurship could be one of the key approaches to reducing the severity of climate change. Pouspourika (2020) stated that green entrepreneurship is the activity of consciously addressing an environmental or social problem or need through the realization of entrepreneurial, innovative ideas with a high level of risk which has a net positive effect on the natural environment while at the same time being financially sustainable.

Brundland report (1987) indicated that sustainable development is the idea that human societies must live and meet their needs without compromising the ability of future generations to meet their own needs. Therefore, the core idea of sustainable development is to work for a society that will exist long-term. Sustainable development focuses on both the imperatives of present and future. It emphasizes the preservation of the environment and natural resources for social and economic equity. Munasinghe and Downing (2003) cited in Bizicova et al. (2007) state that in 2003 the global climate policy turned towards climate change and sustainable development. Integrating climate change with sustainable development leads to many innovative developments. While climate change integrates with sustainable development, Cohen et al. cited in Bizicova et al. (2007), identified two major approaches: integrating sustainable development into climate change policies and integrating climate change into sustainable development policies.

Likewise, Swart et al. (2003) pointed out a strong linkage between climate change and sustainable development. In their paper, they analyze the extent that climate change and sustainable development are strongly correlated and suggest that their correlation not only provides new opportunities but also opens up opportunities for integrated policy development. In 2019, the United Nations declared that climate change is not just one of the seventeen sustainable development goals, it is a threat that contributes to aggravate some other big development challenges such as health, hunger, poverty, and education. Climate change hampers achieving the goals of sustainable development in many ways.

Similarly, Beg et al. (2011) stated that there is a strong connection between climate change and sustainable development in developing countries in such areas as energy efficiency, renewable energy, transport, and sustainable land-use policies. The following table shows the evolving meaning of sustainable development:

Table 1. Definition and Core idea of sustainable development

Author/Publication	Year	Definition and core idea of sustainable development
WCED	1987	Sustainable development is a development that meets the needs of the present without compromising the ability of future generations to meet their own needs.
Pearce et al.	1989	Sustainable development implies a conceptual socio-economic system which ensures the sustainability of goals in the form of real income achievement and improvement of educational standards, health care and the overall quality of life.
Harwood	1990	Sustainable development is an unlimited developing system, where development is focused on achieving greater benefits for humans and more efficient resource use in balance with the environment being required for all humans and all other species.
IUCN, UNDP & WWF	1990	Sustainable development is a process of improving the quality of human life within the framework of the carrying capacity of the sustainable ecosystems.
Lele	1991	Sustainable development is a process of targeted changes that can be repeated forever.
Meadows	1998	Sustainable development is a social construction derived from the long-term evolution of a highly complex system – human population and economic development integrated into ecosystems and biochemical processes of the Earth.

Table 1. (Continued)

Author/Publication	Year	Definition and core idea of sustainable development
PAP/RAC Vander-Merwe & Van-der-Merwe, 1999	1999	Sustainable development is development given by the carrying capacity of an ecosystem. Sustainable development is a programme that changes the economic development process to ensure the basic quality of life, protecting valuable ecosystems and other communities at the same time.
Beck & Wilms	2004	Sustainable development is a powerful global contradiction to the contemporary western culture and lifestyle.
Vare & Scott	2007	Sustainable development is a process of changes, where resources are raised, the direction of investments is determined, the development of technology is focused and the work of different institutions is harmonized. Thus the potential for achieving human needs and desires is increased as well.
Sterling	2010	Sustainable development is a reconciliation of the economy and the environment on a new path of development that will enable the long-term sustainable development of humankind.
Marin et al.	2012	Sustainable development gives a possibility of time unlimited interaction between society, ecosystems, and other living systems without impoverishing the key resources.
Duran et al.	2015	Sustainable development is a development that protects the environment because a sustainable environment enables sustainable development.

Source: Adapted from Klarin (2018:77).

Mensah (2019) stated that the term sustainable development has two core words: 'sustainable' and 'development.' These two terms have been defined from various perspectives. It is also ensuring a balance among three important areas: economic growth, environmental integrity and social well-being. The following figure depicts the complex nexus between the three pillars of sustainable development. These three pillars must be sustainable and mutually balanced in order to achieve sustainable development. However, maintaining a good balance between three pillars is a complex task to achieve.

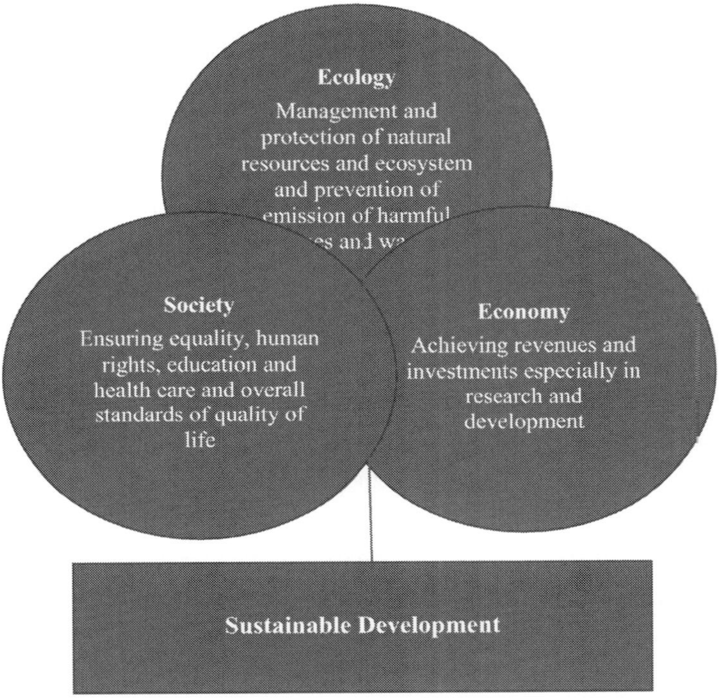

Source: Adapted from Elkington, 1994; OECD, 2000; Dréo, 2006; Brooks, 2013. Cited in Klarin, 2018: 85.

Figure 2. The three pillars of sustainable development Environment.

The above diagram illustrates how three pillars (ecology, economy, and society) of sustainable development are correlated. One pillar affects the other pillar either positively or negatively. Therefore, it is essential to ensure a good balance between these three pillars. According to Bhattacharya et al. (2015: 11) "*sustainable infrastructure is an infrastructure which is socially, economically and environmentally sustainable.*" In their paper they highlight that sustainable infrastructure is inclusive and it meets the needs of the disadvantaged people through infrastructural access and reduction of vulnerabilities to climate change risks. Social perspective of sustainable infrastructure respects human rights. Economic viewpoint of sustainable infrastructure impacts on job outcomes. It increases GDP per capita and include opportunities to build local developer capacity. Economically sustainable infrastructure does not burden government with debt they cannot repay. It also helps poor by reducing tariffs they cannot afford. Finally, environmentally sustainable infrastructure is resilient to climate change. It sheds light on an infrastructure that establishes the foundation for a low carbon economy.

The book argues that environment, climate change and green entrepreneurship are key components to achieve sustainable development. Globally around 1.4 billion people rely on coal and wood to meet basic energy needs. This is not only harmful to the environment but it also causes premature death of women and children. Cooking with coal and wood in interior cooking spaces creates harmful indoor air pollution. Toxic indoor home environments are possibly the most hazardous type of all chemical pollution (Nawaz, 2015: 100). The book also contends that Green entrepreneurship has been regarded as one of the effective climate change mitigation measures to reduce adverse effects of climate change. Green entrepreneurship is the process of wealth creation through optimal use of natural and man-made resources by ensuring the minimisation of greenhouse gas emissions and waste generation to reduce air, water and solid waste pollution, thereby increasing profitability while enhancing an overall positive impact on humanity. The entire world has begun utilizing green entrepreneurship with many countries employing a number of green initiatives.

The U.S. government is creating 4.2 million new jobs in their green economy sector. Many new business models and opportunities have been adopted in Tunisia. South Asian countries are also taking green initiatives for a better world and sustainable economy. Currently, India and Nepal are mapping many green initiatives such as making a green market economy, regenerating power, recycling goods, waste management, sustainable use of natural resources as well as many others. (Sharma& Kushuwala, 2016). United Nations Environmental Program (UNEP, 2009 cited in OECD, 2011) stated that South Korea invested in 2009 79% of its total economic stimulus package in 'green activities' representing almost 7% of its GDP, followed by China and Australia with 34% and 21% of their stimulus packages going to 'green investments,' corresponding to 5.2% and 0.9% of their respective GDP. In this context, the study of green entrepreneurship went from being simply 'fashionable' to being essential for policy guidance. The current scholarship on the issue is quite rich and available in various academic papers, research reports and published books. This book addresses the various prospects and challenges of green entrepreneurship as an effective strategy for combating climate change, which is a complex journey towards sustainable development.

INTERCONNECTION BETWEEN ENVIRONMENT AND GREEN ENTREPRENEURSHIP

Every empirical research has an implicit, if not explicit, research design (Yin, 2009). Research design is basically a strategy or guide that a researcher follows in executing a research project. This book is going to employ a case study method as a strategy of inquiry. Case study is a systematic investigation of a setting or a single subject or a single depositary of documents or a specific event (Berg, 2009:31). The book is based on a study which is going to deal with 'what' and 'how' questions which are similar to Yin's (2003) design. 'What' and 'how' represent explanatory case study questions and our investigation is therefore going to be an explanatory case study. According to Bryman and Bell (2003), researchers in a case study research have freedom to choose between the two approaches: quantitative and qualitative. The author has employed a qualitative research approach. A case study is the most obvious choice which will give this study the strength to explore the prospects and challenges of green entrepreneurship.

UNEP 2011; European Commission 2011, stated that very little progress has been noticed in developed countries despite the extensive adoption of sustainable development by combining economic, environment and social issues simultaneously. It has been argued by Jackson (2009), that economic development policies need to be combined with environmental policies. Smith, and Grin, 2010 and Davies and Mullin,

2011 state that much attention has focused on low carbon initiatives – an attempt to reduce greenhouse gas emissions and so mitigate climate change – and the development of a green economy.

Economic and global transformations have put the global natural environment in danger. The top ten environmental issues that the planet is currently facing are population, climate change, lost biodiversity, the phosphorus and nitrogen cycle, water pollution, ozone layer depletion, ocean acidification, overfishing and deforestation. Rapid development in the economy has been the major cause of environmental loss. Given growing concerns about global warming there is a pressing need to change current business practices which are widely understood to be the main cause of degradation and reach more sustainable modes of operation.

Since the United Nations Conference on Sustainable Development (Rio +20) held in June 2012, the 'green economy' concept has been on everyone's lips. It is regarded as a solution to pressing environmental problems such as ocean, air and soil pollution and, above all, climate change. Today, humankind's ecological footprint is such that we are using 1¼ planets' worth of resources. This means that we are consuming more than the earth can regenerate. This overuse is going to restrict the resources available to future generations in all areas. Natural resources, particularly those used to produce energy, will continue to dwindle until they run out completely, jeopardizing not only economic development, but also national and international peace (Demuth, 2014).

It is widely accepted that green entrepreneurship is the best approach to protecting the environment. As Bangladesh is the land of nature, there is a sparkling opportunity for establishing green entrepreneurship. Recently, researchers in the field of entrepreneurship have increasingly devoted their attention to the interaction between business and entrepreneurship. Although there is no dearth of literature on entrepreneurship, there is hardly any meaningful research conducted on green entrepreneurship or eco entrepreneurship in Bangladesh. Therefore, this research will add to the current stock of knowledge by exploring opportunities and challenges of green entrepreneurship or eco entrepreneurship in Bangladesh. It is hoped that this new research could also be beneficial for academics and researchers who are keen to further explore the field of green entrepreneurship, especially in the context of Bangladesh. It is expected that one of the major policy recommendations will be to increase awareness among green entrepreneurs through seminars, workshops, conferences, and round table discussions. It is also highly important to offer government-run training programs for green entrepreneurs so that they can learn how to integrate environmental, economic, and social concerns into their core businesses. The proposed policy suggestion might be to increase government support for the eco entrepreneurs and implement strong environmental legislation.

SIGNIFICANCE OF THE BOOK

Very little meaningful research has been carried out in the area of green entrepreneurship in relation to climate change and sustainable development in the context of Bangladesh. Therefore, research at this level is essential. Moreover, upon focusing on all possible opportunities, barriers and triggers, this book will add a new dimension to understanding the total scenario of green entrepreneurship in Bangladesh. This book aims to fill this void in the existing body of knowledge. This book is significant for the following reasons:

Firstly, Bangladesh is one of the most climate vulnerable countries in the world and will become more vulnerable day by day as a result of climate change. IPCC (2007b) reported that climate change is defined as changes over time in the averages and variability of surface temperature, precipitation, and wind as well as associated changes in the Earth's atmosphere, oceans and natural water supplies, snow and ice, land surface, ecosystems, and living organisms. Floods, drought and cyclones are likely to become more frequent in coming years. Environmental contamination accelerates climate change gradually. In this context, Voumik and Shah (2014) argued that green entrepreneurship development can be one of the most effective climate change mitigation strategies that a country like

Bangladesh could follow. Therefore, it is timely to write a research-based book on this topic.

Secondly, the Bangladesh Bureau of Statistics (2017) reported that the total population of Bangladesh is around 161.80 million but only 60 million people are employed. If Bangladesh want to increase employment, there is no other way but to increase self-employment. Considering this alarming situation, green entrepreneurship needs to be encouraged by the government and other organizations. Thirdly, like many other countries of the world, our neighbor country India is very interested in green entrepreneurship. Considering the global climate change issue, India has recently adopted many initiatives for promoting the green industry. There are many available research studies focusing on green entrepreneurship in India, but there is little in the way of studies and initiatives surrounding this issue in Bangladesh. Only limited analysis of policy and practice has so far been undertaken on green entrepreneurship in Bangladesh, so thorough research at this level is needed. At the same time, this book will explore the prospects and challenges of green entrepreneurship in relation to climate change in order to propose some policy suggestions for green business in Bangladesh as well as helping to fill current gaps in the research. It is important to mention that existing literature lacks extensive empirical investigation. In last few years there has been growing debate on the topic. There have been several studies conducted in the area of environment and climate change but very little meaningful research has been conducted in the area of climate change and green entrepreneurship.

In a nutshell, the significance of the study is both theoretical and practical. By exploring the various dimensions of green entrepreneurship, it adds value to the current stock of knowledge of entrepreneurship. It is the opportune time to conduct research in this area, as it is anticipated that the findings of the study will be useful for policy makers in the formulation of innovative and useful strategies that will help to promote green entrepreneurship in Bangladesh. It is hoped that this new research could also be beneficial for academics and researchers who are keen to further explore the field of green entrepreneurship, especially in the context of Bangladesh. Dhaka Tribune (2018) reported that Bangladesh has introduced some green policies, but the problem remains in implementing those policies. Therefore, it is high time to do some research on this topic, in order to enforce green policies in Bangladesh.

The Research Terrain

Within this section, the methodological approach is outlined, and the research design and process are discussed. In order to accomplish the objectives of the study, qualitative data collection tools have been employed in gathering information. This section gives a detailed account of the research approach, sources of data, research process, problems encountered during data collection and ethical considerations. Furthermore, the study locations are highlighted, representative sampling methods are explained, and data analysis techniques are demonstrated.

Sources of Data

This book is based on both primary and secondary data. Both primary and secondary sources of data have been used in the study. Primary data has been collected from a number of green entrepreneurs, non-green entrepreneurs, bank officials and Key Informant Interviewees (KII). Data has been validated through crosschecking respondent's views with each other and primary data has also been crosschecked with secondary data. Secondary data has been collected from various publications, journals, books, articles, and websites.

The following diagram reveals the sources of data used in the study.

This book is entirely based on qualitative research methodology. Data were collected through in-depth interviews, focus group discussions and observation. Secondary sources of the research include books, research reports, journals, theses, relevant publications, reports, daily newspapers, and relevant websites.

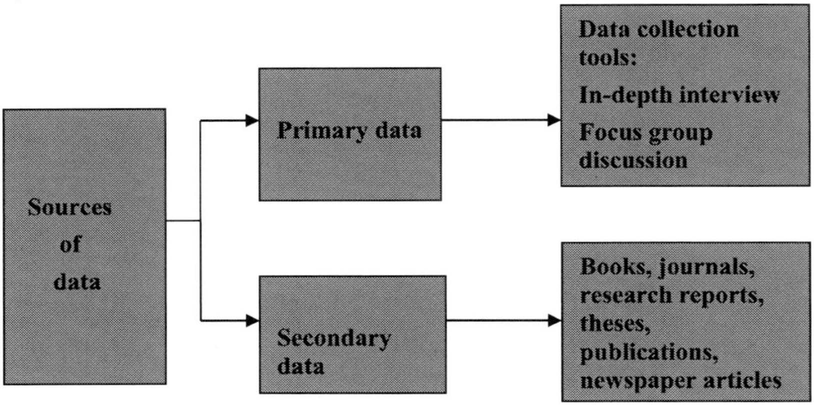

Source: The Author.

Figure 3. Sources of data.

RESEARCH METHOD AND DATA COLLECTION TOOLS

The book follows a case-oriented qualitative research method. In this process, participants are researched in their natural settings and are ideally not controlled by the researcher's preconceived knowledge. The author attempted to collect and interpret data from the perspective of the population under study (Hancock, 1998). As Creswell (1994) points out, there are six assumptions of qualitative research based on Merriam's (1988) work: qualitative researchers are concerned primarily with (i) process, (ii) an interest in understanding how people make sense of their lives, (iii) the researcher as the primary instrument for data collection and analysis, (iv) the involvement and importance of fieldwork, (v) descriptive, observational research that demonstrates an interest in process, meaning and understanding gained through words or pictures, and (vi) inductive qualitative research (Creswell, 1994). Denzin and Lincoln (2000) define qualitative research in the following way:

> Qualitative research refers to a situated activity that locates the observer in the world. It consists of a set of interpretive, material practices that make the world visible. Qualitative research involves an

> interpretive, naturalistic approach to the world. Qualitative research studies things in their natural settings, attempting to make sense of or to interpret phenomena in terms of the meanings people bring to them. (p. 3)

Qualitative methods provide an in-depth explanation of why and how changes arise out of an intervention. "Qualitative methods are useful for understanding the reasons for events described in research or impact assessment" (Simanowitz, 2003: 4). Qualitative tools include semi-structured in-depth interviews, focus group discussions, among others. Punch (1998) states that in-depth and disciplined case studies assist to develop new concepts to explain some aspects of what has been studied (cited in Nawaz, 2007: 79). Holtzhausen (2001) argues that triangulation is a powerful tool that strengthens the qualitative research design. This study followed the triangulation method because different qualitative approaches in the data collection were used, involving multi-perspectives coming from both green entrepreneurs and key informant interviews (KII). Multiple qualitative data collection tools such as in-depth interviews, focus group discussions and participant observation were applied in order to ensure the validity of the research study. Triangulation is a safety net for any qualitative research. It establishes the validity of the data collected and diminishes any biases the author may have as a result of their prior knowledge or preconceived assumptions and theories (Creswell, 2012).

In summary, as the present study follows a qualitative strategy, the process used inductive reasoning, because it was exploratory and moved to a broader generalization and theories, based on the on-going construction of reality by both the participants and the researcher. The ideas and actions of the participants and the method of construction of their realities were given the highest priority. The study emphasises 'words' rather than 'numbers' and generates theories rather than generalising populations. Moreover, the research focused on the process—why and how things happen rather than the outcome. I tried to engage in the whole process of research to have an in-depth perception of a smaller number of cases. Therefore, importance was given to 'depth' rather than 'breadth' and to the participants' real world. My interpretive epistemological position drove my choice of methodology, guiding me to choose inductive research and qualitative data collection tools to conduct the study.

Sampling and Recruitment Process

Table 2. Total sample size for this research study is as follows:

Categories of respondents	Total No
Green Entrepreneurs	15
Non-Green Entrepreneurs	10
Bank Officials	5
KII	5
Total	45

Source: Fieldwork conducted by the Author.

It is argued by some that too large a sample may involve huge cost, workforce, material and time while too small a sample may invalidate the result. Around 30 respondents are often considered the bare minimum for undertaking statistical data analysis. On the other hand, many researchers regard 100 cases as the necessary standard for achieving reliable results (Fisher et al. 1991 cited in Islam, 2008: 115). However, this research did not require a very large sample size as a case-oriented qualitative research approach was applied. The sample was selected for the purpose of the study not on the basis of a representative population. As Yin (1994) argues, the evaluation of case studies should be based on the theoretical construct, not on the size of the sample, as is done in conventional quantitative strategies. The main purpose was to have a complete in-depth understanding of the case. I recruited the participants in five stages. Photographs were also taken of specific sites and situations to demonstrate the current situation of green entrepreneurship in Bangladesh. As discussed previously, the book is based on a study used a combination of in-depth interviews, focus group discussions and observation as methods of data collection. The following process was undertaken for collecting data from the selected samples.

In-Depth Interviews

Primary data consistent with the research objectives and questions was collected through in-depth interviews. This is a verbal technique for obtaining data directly from the primary source (Aminuzzaman, 1991: 82). In-depth interviews were conducted with fifteen green entrepreneurs. Green entrepreneurs were chosen through purposive network sampling for detailed interviewing. Purposive sample sizes are often determined on the basis of theoretical saturation (Auerbach & Silverstein, 2003: 102). Theoretical saturation is the point in data collection when new data no longer brings additional insights to the research questions.

In Stage one, ten entrepreneurs were also chosen who were not adopting green methods. Fifteen green entrepreneurs were chosen through purposive network sampling. All these entrepreneurs were interviewed at their workplace. They were asked to answer a series of open-ended questions. The main aim of the in-depth interview was to collect qualitative data to examine the challenges and opportunities of green entrepreneurship in Bangladesh.

In stage two, five bank officials were interviewed from Bangladesh Bank, Sonali Bank and Eastern Bank, to understand about the activities of green financing in Bangladesh. In stage three, five key informant interviewees were chosen to understand their views regarding the opportunities and barriers of green entrepreneurship in Bangladesh.

In stage three, a number of focus group discussions (FGDs) were arranged with green entrepreneurs. The FGDs were conducted to understand and compare the views of participants about the challenges and prospects of adopting green entrepreneurship. This method was very useful for collecting more in-depth information since people in Bangladesh prefer collective rather than individual discussions. The respondents from the first phase of the study were invited and subsequently agreed to participate in the second phase. Three FGDs were conducted with fifteen green entrepreneurs to explore and understand their views about green entrepreneurship. Two FGDs were conducted with non-green entrepreneurs. In order to help the participants to focus on the questions and lessen the temptation to engage in personal discussions with one another, only five respondents were included in each group. The author met with these small groups of respondents to address specific general questions. The author also tried to keep the discussion within the research framework.

In stage four observation refers to a technique which operates by systematically selecting, watching, listening, reading, and recording behavior and characteristics of living beings, objects or phenomena (Islam, 2008: 171). The observation method was also used for collecting data from the respondents. This method was very useful for collecting actual data about respondents' behavior and attitudes during the process of my interviewing. Qualitative methods have been applied in understanding the whole context of green entrepreneurship in Bangladesh. Using a qualitative approach, the author will be able to understand different realities, and the way that those realities are perceived by eco entrepreneurs. Key documents have been studied and used in the study. Case stories were also obtained through face-to-face, in-depth interviews with the research respondents. Qualitative data analysis techniques have been used for the proposed study.

Selection of Case

The selection of cases is the most important aspect of qualitative studies and crucial for later analysis. Qualitative sampling, unlike quantitative sampling, tends to be purposive rather than random. The choice of informants, episodes and interactions was being driven by a conceptual question, not by concern for "representativeness" (Miles & Huberman, 1994). Sampling in qualitative research is usually not wholly pre-specified, but it evolves once fieldwork begins.

The Research Processes

The study areas were selected with great care so as to maximize the quality of the results. During her field visit, the author clearly described the purpose of the interviews to the respondents. She informed them that the information they shared with her would only be used for her academic book project. She also explained that she had no connections with any other organisations. She endeavoured to develop a trusting relationship with them by using a friendly and simple attitude.

The Process of Data Collection

The core research has been formed by project director and one research associate. The team worked rigorously to collect a wide range of data from the respondents. A number of interested students were selected as data collectors. Both female and male data collectors were selected for the study. Students studying Public Administration got priority to work as data collectors in order to help the project researching study objectives of that course. Before going to the field, they received an orientation about the study and instruments of data collection.

Challenges of Data Collection

Collecting primary data for social research from rural areas of developing countries such as Bangladesh, is a very critical and problematic task. In this case the author had to face numerous challenges. Before going to the field, she made phone calls to offices of respective green entrepreneurs, but they asked her to send a formal letter by email asking their permission to conduct research with them. She also needed permission to visit the industries. For seeking permission, she had to exchange further emails confirming that the information that she would gather from the field would be used for academic purposes only.

As soon as the author arrived in Rajshahi, she made a courtesy call to the offices of 'Ifat Bio CNG Gas Plant ltd, Nabil Group and Grameen Feed. They all were very polite while talking over the telephone, however, when she visited the field offices, the officials were unco-operative until the formal permission letter was produced. When she met with the various entrepreneurs, they seemed very co-operative. Despite introducing herself with the letter of introduction stating her identity and purpose, some had the impression that she could do something for them. They were mostly interested in talking about their problems and their needs which were not related to the research questions. The author had to constantly remind them of the reasons for talking with them and this resulted in some losing interest and quitting the interview session.

Ethical Consideration for Data Collection

Ethical issues are a very important consideration when undertaking human based research studies. At any phase of research, subjects might be exposed to physical, emotional, or psychological harm that may constitute the violation of human rights. Therefore, ethical considerations in social research must be addressed in order to prevent such harms (Nachmias & Nachmias, 1996).

Violating respondents' privacy may create a threat to their integrity and may cause emotional and psychological anxiety and should be avoided by any means. Participants were assured that no identifying information would be published in the report unless prior approval was given. Their privacy would also be maintained and the confidentiality of any personal information shared by them would be respected. The author clearly described the purpose of the interviews, explaining that their participation would not affect them in any way and she also informed them that the information would be used for academic purpose only. All recorded interviews were retained by the author and were not made available for any other purpose. All interviews with the participants were audio-taped with their permission and later transcribed and then translated into English.

Photographs were taken only with participants' full consent. Any identifying information was removed and the typed-up files were stored on a password-protected computer to which was only accessible by the author.

The participants were anonymous, and the author endeavoured to always maintain their anonymity. To address this issue, throughout the research, the author used pseudonyms to describe each respondent's narratives in order to ensure their anonymity. As gossip could lead to negative consequences, the author was particularly careful when managing the focus groups. The author always tried to minimize risk to participants. As an academic in Bangladesh, she has had previous experience conducting research with similar participants, and therefore had the background knowledge and familiarity to address any problems she encountered. For minimising any possible perception of pressure, the participants were not invited to participate in stage two until after they had completed stage one. This process occurred through self-selection. After the first introductory meeting, the author allowed potential participants plenty of time to consider their participation in the research. Her intention was to recruit only those respondents who were interested. Fortunately, all participants of stage one were willing to participate in stage two.

The author informed the respondents that participation in the research was absolutely voluntary and that they could refuse to answer any questions at any time. She also gave them full authority to withdraw from the interview at any time without effects or consequences. An important ethical consideration was to ensure that participants were well informed about the objectives, benefits, and potential risks of their involvement in the book project. It was important to inform them about the main elements of the book project so that they could make autonomous decisions as to whether or not they wished to participate. The author met with the respondents directly to gain their consent for their participation and she recorded informed consent in various ways. According to Nachmias and Nachmias (1996:83), "informed consent is the procedure by which individuals choose whether to participate in an investigation after being informed of facts that are likely to affect their decisions." She also informed the participants about tentative time, the research process, and the role they would play. During her field work, she constantly reminded the respondents that their involvement in the research was voluntary and they had no obligation to participate.

Profile of Research Site

This section provides the geographical details of the study area, infrastructure and social composition of the study site. As indicated earlier, the study was carried out in the Rajshahi District of Bangladesh. Rajshahi district is situated in the north-western part of Bangladesh. Originally it was a subdivision of the previous greater Rajshahi division. In 1972 the subdivision was upgraded to a district. The following map depicts the study area.

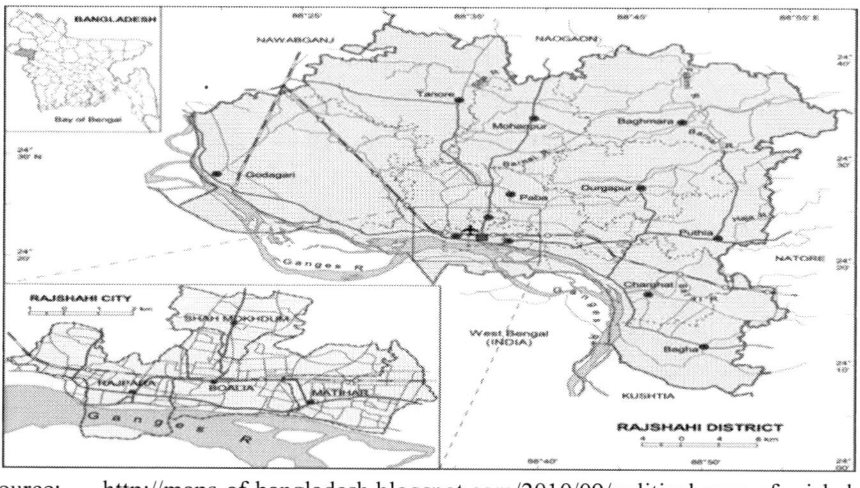

Source: http://maps-of-bangladesh.blogspot.com/2010/09/political-map-of-rajshahi-district.html.

Map 1. Map of Bangladesh and Rajshahi District.

Separate districts namely Maldaha, Bogra, Pabna, Natore and Nawabganj were established by dividing the broader Rajshahi district. Rajshahi has one city corporation, 13 *Upazila*s (four in metropolitan areas),

71 *Unions*, 14 *Pauroshavas*, 1,718 *Mousers* and 1,914 villages. Rajshahi district consists of nine Upazillas (administrative regions acting as sub-units of districts): Bagha, Bagmara, Charghat, Durgapur, Godagari, Mohanpur, Paba, Puthia and Tanore. The river Padma flows through the periphery of Rajshahi city. The area of Rajshahi district is 2,407.01 square kilometres. The predominantly agriculture-based district comprises a population of 2,300,000 people. The percentage of male and female is 50.66% and 49.34% respectively. Muslims make up 93% of the total population with the remainder being Hindu, Christians, and other religions. Agriculture is the main occupation for 38.73% of the population. 23.64% are agricultural labourers, 12.44% are involved in commerce, with the remainder engaged in other occupations. There are 559 public and 421 private educational institutions in the district (http://www.dc rajshahi.gov.bd).

Data Processing and Analysis

As a first step, all respondents' interviews were recorded. The second step involved manually transcribing, translating, and organizing the information that was collected. As a means of thematic analysis, interviews were categorised into different subthemes under a main theme and were then analysed for meaning. NVIVO was used to store and examine data. For data processing and analysing, thematic and conversational analysis were adopted following Denzin and Lincoln's (2000) method, especially for identifying themes. Themes are abstract constructs identified by researchers before, during and after data collection. Many themes were also identified during the literature review, as well as while conducting the fieldwork and collecting the field data. Themes were constructed following Williams et al. (1990) and Miles and Huberman (1994, cited in Denzin & Lincoln, 2000: 781), with general themes arising from reading the literature and subthemes arising through the progression of the study.

To determine the final results, all themes were reviewed, critically analysed and connected to the research questions (Miles & Huberman,

1994:12). A codebook was produced to detail the descriptions of each code and record the inclusion and exclusion criteria and exemplars of the text for each theme. For ease of application and recall, some codes applied are words and some are numbers. While coding, the author was reminded of Miles' (1979, cited in Denzin & Lincoln, 2000: 781) discussion about coding and code books where it was reported that codebooks are simply organized lists of codes and that coding is supposed to be data reduction, not proliferation. Miles also argued that codes themselves are mnemonic devices used to identify or mark the specific themes in a text. The coding process allows the identification of themes by regrouping concepts with shared meaning (Bazeley, 2007: 71).

To manage the qualitative data, tabulation was prepared. Every interview was analyzed separately and the findings were re-checked against the views of other respondents. Data display was supported by narratives from the participants. All respondents were informed about the use and dissemination of the data. Through the interviews and data entry, a sound understanding of the research issues and problems was obtained. Qualitative information was used to interpret the findings and to illustrate and support the explanations.

OUTLINE OF THE BOOK

Chapter one reveals the framework of the book. This chapter states the research problem, sets research objectives, and formulates the research questions. The chapter also focuses on the scope and limitations of the study and sheds light on the justification for undertaking research in this particular area. Chapter one also discusses the research design and process. It includes a detailed account of the sources of data, research methods, research design, research sites, sampling methods, research processes, problems encountered during data collection and ethical considerations. The respondents were chosen through a purposive network sampling method. Both primary and secondary sources of data were used in the study. Primary data were collected from green and non-green entrepreneurs, bank officials and key informant interviewees. Data were validated through crosschecking one respondent's views with the other and primary data were also crosschecked with secondary data. Secondary data were collected from various publications, books, articles and websites. When collecting data, my interpretive epistemological position guided my choice of applying qualitative methods such as in-depth interviews, focus group discussion and observation. Case stories were also obtained through in-depth interviews with green entrepreneurs. Data were validated by applying triangulation data collection methods and this chapter also provides details of the data analysis techniques that were used.

Chapter two highlights the various key concepts of the research study that have been applied throughout the book. The chapter sheds light on various key concepts in order to draw a theoretical foundation for the study. It demonstrates the theoretical foundation climate change and green entrepreneurship. This chapter also highlights a range of literature relevant to the study that identifies a gap in current understanding. This provides an insight into the significance of my research. Finally, this chapter presents the conceptual model used to guide the study. Various concepts of the model have also been operationalised in this chapter.

Chapter three highlights detailed description and background of sustainable development. It discusses the significance of sustainable development goals that have to be achieved by developing countries by 2030. It sheds light on the extend to what and how sustainable development goals addresses environment, climate change and green entrepreneurship. Chapter four shows global and national context of climate change and green entrepreneurship. The chapter begins with the global perspective which is followed by Asian perspective. Then it concludes with a brief description about Nationally Appropriate Mitigation Action (NAMA) to understand the existing gap at policy level. Lastly Chapter five highlights findings and analysis which is followed by some policy suggestions.

Chapter 2

THE THEORETICAL FOUNDATION OF CLIMATE CHANGE AND GREEN ENTREPRENEURSHIP

This chapter highlights the theoretical foundation of climate change and green entrepreneurship. This book argues that green entrepreneurship has been considered as one of the effective strategies for mitigating climate change. The Intergovernmental Panel on Climate Change (2007) defines climate change as any change in climate over time, whether due to natural variability or as a result of human activity. It more than merely changes in weather. Changes in climate refer to changes in the means and variability of, for instance, temperature, precipitation, and wind over the course of months to millions of years. More elaborately, climate change refers to changes in atmosphere, cryosphere, land surface and biosphere. Atmosphere includes the gaseous cover surrounding the earth, hydrosphere refers to water on the surface of the earth in both the ocean and land surfaces, while the biosphere includes ecosystems and organisms living in the atmosphere, land and oceans.

The negative impacts of climate change include heat-related deaths and illnesses, heat-related impacts on livestock, wildlife, and, agriculture, extended range of some pest and disease vectors, increase in the number of floods, land and mudslides, and soil erosion. Further impacts are increasing summer drying and drought which in turn causes a decrease in crop yields and water resource quality and quantity, and an increased risk of forest fire. Other negative impacts are an increase in tropical cyclone wind and precipitation intensities leading to increased risk to human health, risk of infectious disease epidemics, coastal erosion and damage to coastal infrastructure, and damage to coastal ecosystems. The IPCC (2007) report outlines several domains in which consequences are occurring or are expected from climate change. The following figure depicts the schematic framework of anthropogenic climate change drivers, impacts and responses.

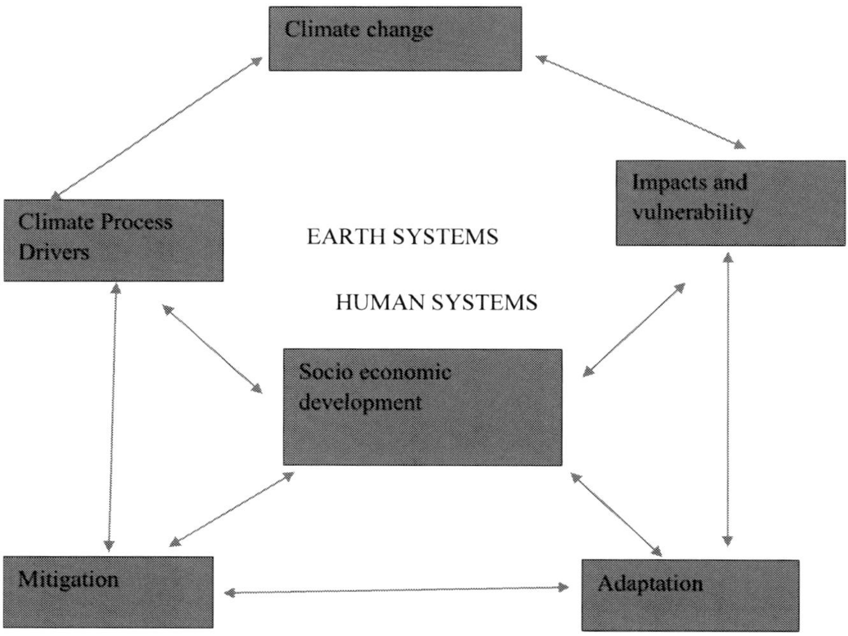

Source: Adapted from IPCC, 2007a, A report of the American Psychological Association, Taskforce on the interface between psychology and global climate change.

Figure 4. Schematic framework of anthropogenic climate change drivers, impacts, and responses.

The above figure depicts impacts, vulnerability and responses to climate change. Here climate change includes extreme events, sea level rises, temperature change and precipitation change that effects on ecosystem (Food Security Water Resources Settlements & Society Human Health). Impacts and vulnerabilities to climate change classified as climate change mitigation and climate change adaptation. Climate change mitigation includes human interventions to reduce drivers of climate systems and climate change adaptation includes managing natural and human systems in response to variations in the climate and their effects. Finally, climate process and drivers are affected by socio-economic development. Climate change mitigation measures include industrial processes, electric public transport and increased carbon tax, all of which can lessen climate change.

The book argues that climate change mitigation indicates all efforts to reduce the greenhouse gas for example reducing deforestation, low carbon energy generation and promoting renewable energies change whereas adaptation measures are based on reducing vulnerability to the effects of climate change. In terms of adaptation measures, there are several actions that help in reducing vulnerability to the consequences of climate change. These are more secure facility locations and infrastructures, landscape restoration (natural landscape) and reforestation, flexible and diverse farming to be prepared for natural catastrophes, research and development on possible catastrophes, temperature behavior as well as the implementation of preventive and precautionary measures (evacuation plans, health issues, etc.) that help reducing vulnerability to the consequences of climate change.

Swathi (Cited in Rao, 2017) argued that climate change is considered as a critical comprehensive global challenge. The adverse effects of climate change impact a wide variety of different spheres including agriculture, food security, rising sea levels, coastal zone erosion and they enhance the intensity of different types of natural disasters. Rao also pointed out that climate change happens due to increased greenhouse gas emissions caused by fossil fuel use and increases in carbon dioxide levels in the atmosphere. This problem is then amplified further by the rapid growth of deforestation; the level of carbon dioxide in the atmosphere is being continuously increased because the decrease in residual forests cannot absorb gases that are harmful for the atmosphere. These harmful gases increase the overall greenhouse gas amount in the atmosphere which contributes to global warming.

Natural processes can also contribute to climate change. Natural processes and factors encompass changes of sun's intensity, volcanic eruptions as well as changes in ocean current circulation. From the above discussion it is evident that climate change can result from both natural and man-made factors. Human made factor incorporates human activities responsible for greenhouse gas emissions (IPCC, 2007). The latest evaluation of IPCC is more extensive and shows a detailed overview of climate change mitigation strategies and implementation obstacles. It also incorporates mitigation technologies, execution barriers and policy alternatives for some sectors including transport, industry, agriculture, forestry and waste management.

ENTREPRENEURSHIP AND GREEN ENTREPRENEURSHIP

Before discussing green entrepreneurship, it is fundamental to define entrepreneurship. An entrepreneur can be defined as one who initiates and establishes an economic activity or enterprise. Entrepreneurship thus refers to the general trend of setting up new enterprises in a society (Begum, 1993). The International Labor Organization (ILO, 1984, cited in Islam and Aktaruzzaman, 2011) states that an entrepreneur must have a set of characteristics that typically includes self-confidence, risk taking, leadership, originality and are future and result oriented. Khanka (2002) referred to entrepreneurs as those who innovate, imitate or adopt a business activity. Given that entrepreneurship is the set of activities performed by an entrepreneur, it could be argued that being an entrepreneur precedes entrepreneurship. In any case, the entrepreneurial definitions described above highlight the aspects of risk-taking, innovating and resource organizing.

Table 3. Description of Entrepreneurship

Authors	Year	Description of Entrepreneurship
Cantillon	1730	Entrepreneurship is self-employment with an uncertain return. Cantillon recognized three classes of economic agents: landowners, entrepreneurs and employees.
Baudeau	1767	Entrepreneurship is innovative management
Say	1803	Described entrepreneur as an extraordinarily talented manager
Menger Marshall	1871 1890	Differentiated entrepreneurial decision-making into four sequential stages Hinted at a distinction between entrepreneurs and managers
Shumpeter	1910	Described the entrepreneur as an innovator, carrying out new combinations
Knight		The courage to bear uncertainty
Penrose		Managerial capacities should be distinguished from entrepreneurial capacity Identifying and exploiting opportunistic ideas for expansion of smaller enterprises is the essential aspect of entrepreneurship
Lebenstein	1970	The reduction of organizational inefficiency and at the same time reversal of organizational entropy
Adam	1989	An entrepreneur is a person whose business actions make him/her a leader in the economic world especially when results in industrial growth or technical advanced
Kao	1993	Process of doing something new and something different for the purpose of creating wealth for the individual and adding value to society
Thornton & Flynn;	2003	Discovery and exploitation of opportunities and creation of new organization, which occurs as a context-dependent social and economic process

Source: Adapted from Zahedi and Otterpohl (2015:197).

Burzyńska et al. (2018) argued that entrepreneurs are businesspeople who realize new business opportunities by taking risks and introducing innovations and new ideas that determine socio-economic development. They further stated that the concept of entrepreneurship is immanently linked to the development of the economy, which is achieved through establishment of new enterprises, their successive development, creation of new resources, markets, technologies, products and services, processes and transactions, thus contributing to creation of value. In addition to looking at concepts and characteristics of entrepreneurship, it is also possible to look at the main features of entrepreneurship, which are given as: (i) being an economic agent, (ii) being creative, (iii) taking and bearing risk, (iv) being innovative, and (v) being dynamic. The following are the features of entrepreneurship:

- Being an Economic Agent: Entrepreneurship is mostly an economic function because it involves the formation of an enterprise. It is basically concerned with the production and distribution of services.
- Being Creative: Entrepreneurship is a resourceful response to changes in the environment. It involves innovation or introduction of something new and better. An entrepreneur is a change agent.
- Taking and Bearing Risk: Risk is an inherent, intrinsic and inseparable element of entrepreneurship. An entrepreneur assumes the uncertainty of future. In the pursuit of profits, there is every possibility of loss.
- Being Innovative: Entrepreneurship is an innovative function as it involves doing things in new and better ways. Innovation may take in several forms e.g., a new product, a new source raw material, a new market, a new method of production etc.
- Being Dynamic: Entrepreneurship is a dynamic function. Entrepreneurs thrive on changes in the environment that bring useful opportunities for business. Entrepreneurs always need to take different dynamic decisions. Thus, entrepreneurship is a multi-dimensional concept. It is both an art as well as a science, but it is more an art than science. There are very few ground rules or principles that can be used to create and run business enterprises in a fast-changing and heterogeneous environment.

There are various types of entrepreneurship. The following table illustrates typology of entrepreneurship.

The author conceptualizes entrepreneurship as the process of being an entrepreneur. Entrepreneurship can be considered as what entrepreneurs do. It is a process of involving various actions to be undertaken to establish an enterprise. Some have referred to entrepreneurship as risk bearing while others view it as innovating, and others even consider it to be thrill-seeking (Khanka, 2002). Entrepreneurship is the key to small business development. A country may have resources but the problem of development is to put these resources into the process of production. This can only be made possible through initiatives by people or organizations (Hossain and Rahman, 1999). Entrepreneurship plays a vital role in the growth and development of any country, especially of a poverty-ridden country like Bangladesh.

Table 4. The following table shows different types of entrepreneurship

Type of entrepreneurship	**Core Motivation**
Ecoprenurship	Contribute to solving environmental problem and create economic value
Social entrepreneurship	Contribute to solving societal problem and create value for society.
Institutional entrepreneurship	Contribute to changing regulatory, societal and market institutions
Sustainable entrepreneurship	Contribute to solving societal and environmental problems through the realization of a successful business

Source: Zahedi and Otterpohl (2015:199).

The development of a country depends not only on the human resources but also on the proper utilization of all the resources. Entrepreneurship development is a powerful instrument to activate the forces of socio-economic development of a country. Development of entrepreneurship can solve acute problems like unemployment, underemployment, disguised employment etc. (Saha, 1995 cited in Nawaz). Entrepreneurship thus refers to the general trend of setting up new enterprises in a society (Begum, 1993). Aktaruddin (1999) described personal attributes as being a key factor for entrepreneurial success or failure, while Aktaruddin (2000) focused on the social-economic background of the entrepreneurs. Afrin et al. (2010) identified the factors related to the development of entrepreneurship among the rural borrowers.

Using a multivariate analysis, they showed that financial management skills and the group identity of the borrowers have a significant relationship with the development of rural entrepreneurship. Habibullah (1987) stated that training is an effective tool for entrepreneurship development in Bangladesh. Rahman et al. (2000) critically appraised the impact of different programs and policies relating to entrepreneurship and small enterprises. Rahman (1999) emphasized the importance of project viability, collateral and entrepreneurial evaluation at the project appraisal level of banks. Saleh (1995) found inadequate cash flows, marketing deficits and discriminating treatment from supportive service agencies to create obstacles to entrepreneurship development.

GREEN ENTREPRENEURSHIP

Green entrepreneurship is a combination of environment and entrepreneurship. It attempts to protect the natural environment from all adverse effects. Other objectives of green entrepreneurship include recycling waste products and increasing the use of renewable energy sources to mitigate climate change (Dubash, 2020). The declaration of Stockholm (1972) is the most significant move focusing on global environmental issues. In 1987, the World Commission on Environment and Development (WCED) published a report named ‘Our common future.’ The report came to be known as Brundtland Report.

This book gives importance to sustainable development progress which is considered as a constitution of the green economy. In 1992, United Nations organized another big conference in Rio which created an action plan named “Agenda 21.” Agenda 21 will be elaborately described in next chapter. It is a noteworthy footstep to address global climate change that has led to the Kyoto Protocol. In September 2000, the Millennium Meeting carried out by the UN was another cornerstone along with Kyoto Protocol for focusing on the goals of sustainable green entrepreneurial activities (Kusat, 2013). The Kyoto Protocol refers to an international treaty which has targeted thirty-seven industrialized countries in an attempt to reduce greenhouse gas emissions. While convention inspires industrialized countries to stabilize greenhouse gas emissions, protocol commits them to do so (UNFCC, 2008). The protocol will be elaborately described in following chapter.

Green economy refers to the economic activities system that includes the production, consumption and distribution of goods and services that increase the quality of people's life (Uslu, 2015). The Brundland report (1987) stated that the green economy focuses on current development by protecting the need for future generations. According to OECD (2011) "'Green Development' and 'Green Economy' should more focus on the investment and consumption of some goods and services that contribute to environmental improvement." The first attempt to define the terms related to green entrepreneurs and green entrepreneurship emerged in the early 1990's. Bennet, Berle and Blue were the first researchers who combined entrepreneurship with the environment. The core idea was to promote entrepreneurship through protecting the environment in a sustainable way. Nowadays, green entrepreneurship has been associated with innovation. It has been argued by Walas and Trebacz that entrepreneurship does not exist without innovation. Nikolaou (2018) identifies four categories of green entrepreneurs on the basis of institutional theory and resource-based view.

- *Innovative:* green entrepreneurs who try to identify new opportunities to overcome institutional pressures.
- *Visionary:* green entrepreneurs are those who invest in green entrepreneurship from idealistic reasons and desire to make positive changes in the world.
- *Strategic:* green entrepreneurs who invest in green entrepreneurship in order to identify new opportunities in the market.

In this book, I have chosen both innovative and strategic green entrepreneurs who are trying to identify new opportunities to earn profit and protect the environment in a sustainable way. The following table demonstrates definitions of green entrepreneurship and related concepts.

Table 5. Definitions of green entrepreneurship and related concepts

Green entrepreneurial activity	The process of identifying, evaluating and seizing entrepreneurial opportunities that minimize a venture's impact on the natural environment and therefore create benefits for society as a whole and for local communities.
Green entrepreneurship	Green entrepreneurs engage in "a kind of social activity that aims at protecting and preserving the natural environment."
Environmental orientation	Importance of environmental issues facing firms by mainstreaming green product strategies.
Green logic	Part of a complex institutional environment, facing a sharing platform, alongside the social and economic logic."
Green entrepreneurs	"Those who start businesses based on the principle of sustainability with strong underlying green values and who sell green products or services."

Source: Alwakid et. al. (2020).

Box 1. Key characteristics of green entrepreneurs

- Lack of risk aversion. They take new business opportunities and engage in undertakings which usually involve very high risks;
- Strong internal motivation. Their activities have a generally positive impact on the environment and on economic stability, and they consciously seek to secure a more sustainable future;
- Concentration on a combination of business and a positive impact on the environment while at the same time gaining significant financial benefits;
- Awareness of creating a more sustainable future

Source: Farinelli and Bottini (2011).

There are many terms available to describe green entrepreneurship such as eco- entrepreneurship, environmental entrepreneurship, sustainable entrepreneurship, ecological entrepreneurship, enviro-preneurship or sustainopreneurship. Green entrepreneurship is also termed as green business. A green business is the activity of buying and selling goods and services to make profit as well as protect environment. Green business is both profit and nature oriented which must be financially viable. Green entrepreneurship promotes sustainable development where development is considered as meeting the needs of the present without compromising the ability of future generations to meet their own needs (Brundtland, 1987). Farinelli and Bottini (2011) identified key characteristics of green entrepreneurs.

A green entrepreneur refers to an entrepreneur who is either making a business 'green' or entering into 'green business.' Here one is a process approach and the other is output approach. While defining green entrepreneurship, some authors incorporate ethical, social, or environmental motivation in green entrepreneurial activity. The following illustrate the broad range of possible interpretations of the concept.

Table 6. Various Interpretations of the concept

Isaak (2005)	An Eco entrepreneur is a person who seeks to transform a sector of the economy towards sustainability by starting a business in that sector with a green design, with green processes and with a life-long commitment to sustainability in everything that is said and done.
Volery (2002)	There exist two types of Eco entrepreneurs: 1) "environment-conscious entrepreneurs", are individuals who develop any kind of innovation (product, service, process) that either reduces resource use and impacts or improves cost efficiencies while moving towards a zero-waste target 2) "green entrepreneurs", are those who are both aware of environmental issues and whose business venture is in the environmental marketplace. Such entrepreneurs pursue environmental-centered opportunities which show good profit prospects.
Anderson (1998)	Both Entrepreneurship and Environmentalism are based on a perception of value. The attitudes which inform environmental concern create areas of value that can be exploited entrepreneurially. "Environmental Entrepreneurs" not only recognize opportunity but construct real organizations to capture and fix change in society.

Source: Adapted from OECD, 2011:25.

Fossil fuels such as oil, kerosene and natural gas are finite and burning fossil fuels contributes to greenhouse gas proliferation in the atmosphere and compounds the negative impacts of climate change. Green entrepreneurship, or green business, might be the solution to the negative effects of climate change. Green business consists of balancing between financial, social and environmental aspects as a fundamental business strategy. In other words, there is no single definition of green business.

However, green business refers to a model that addresses an environmental issue. It also improves energy or resource efficiency, reduces greenhouse gas emissions and/or decreases waste or pollution. Green business includes the following:

Box 2. Green Business includes the following characteristics

- It incorporates the basic principles of sustainability.
- It should distribute benefits equally across the value chain.
- Maximizes the social benefits of the business by incorporating marginalized groups.
- It looks for every opportunity to increase the social benefits of the business.
- Green business supplies environment friendly products and services that replace non-green products.
- It focuses on sustainability by reducing pollution.
- Green business focuses on resource (i.e. energy, water and materials) utilization.
- Green business emphasizes on replanting trees, enhancing soil fertility and using renewable energy.
- It adopts environmental principles in its business operations.

Source: Brundtland, 1987.

Social Entrepreneurship vs Green Entrepreneurship

The following are the three characteristics of social entrepreneurship:

- Integrating environmental, economic & social axis in core business.
- Innovative solutions to the way goods and services are produced & consumed.
- Scaling-up of the business model contributes to greening of the economy.

Green entrepreneurship can be distinguished from the green social entrepreneur's community. As mentioned before, a green entrepreneur can be either making a business 'green' or simply entering a 'green business'

(OECD, 2011). In this book, the author has chosen the entrepreneurs who have simply entered in 'green businesses. In other words, green entrepreneurship seeks for ecological sustainability, while green social entrepreneur's community tries to make changes in the relationship between the economy, ecology and society towards a sustainable approach (Zahedi and Otterpohl, 2015).

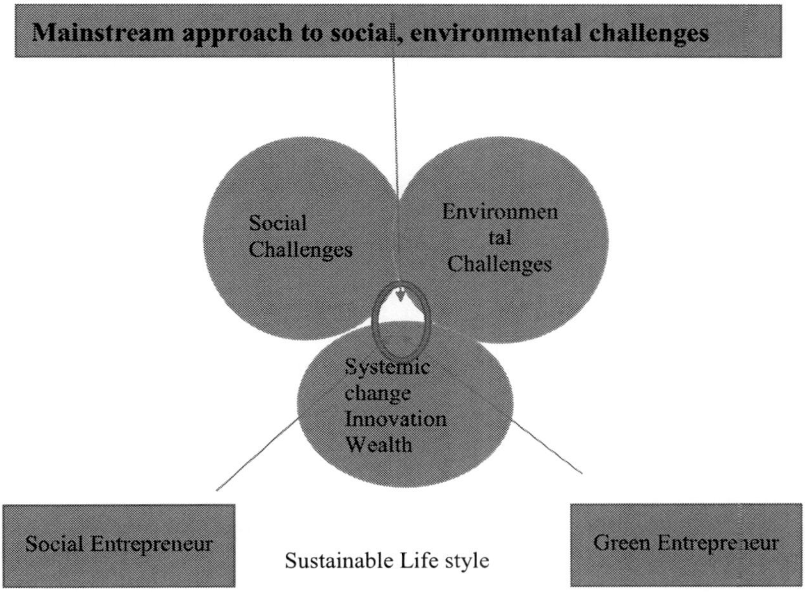

Source: Allam (2011).

Figure 5. Social Entrepreneurship VS Green Entrepreneurship.

The intellectual enterprise of green entrepreneurship began in the late 1970s in the United States of America. Since then, it has expanded all over the world. Gevrenova (2015) stated 'green entrepreneurship' could be defined as a new type of commerce which is equally profit and nature-orientated and might possibly solve ecological problems throughout its business activity. The following table illustrates the broad range of possible interpretations of the concept:

In this book, the definition of green entrepreneurship adopted is based on the type of output produced by firms. The concept of green entrepreneurship is relatively new and has been receiving growing attention since 1990. Green entrepreneurship emerges as a mutual product of environmental protection and entrepreneurship. The core idea of green entrepreneurship is to provide positive environmental outcomes when producing goods and services (Uslu et. al, 2015). Green entrepreneurship is associated with the green economy. It is an attempt which is reliant on both the economic and social conditions. One of the most important aims of green entrepreneurship is to prevent the natural environment from all adverse effects. The other main targets of green entrepreneurship are as follows:

- recycle waste products,
- increase the use of renewable energy sources,
- and to make organic agriculture and farming.

The United Nations International Labor Organization defined green entrepreneurship from two perspectives: one is relating to output (products and services), as the other is the process of an economic activity. Green entrepreneurship can happen in two ways: firstly, existing entrepreneurs can enter into an explicitly 'green business' who provides environmentally friendly products and services through waste management. Consecutively, green entrepreneurs can provide their products and services through environmentally friendly processes such as eco-tourism. Green entrepreneurship concerns individuals and organizations engaged in entrepreneurial activities that create environmental benefits by offering green final products or services. It also refers to an organization's efforts at designing, promoting, pricing and distributing products that will not harm the environment. As resources are limited and human wants are unlimited, it is essential for the marketers to utilize the resources efficiently without waste as well as to achieve the enterprise objective. Green marketing is therefore an essential tool. There is growing interest among consumers all over the world regarding protection of environment. Worldwide evidence indicates that people are concerned about the environment and are changing their behavior. As a result of this, green marketing has emerged which speaks for the growing market for sustainable and socially responsible products and services (Mishra & Sharma 2010).

To sum up, green entrepreneurship is one of the effective strategies for mitigating climate change. The author argues that if green entrepreneurship is adopted, it will surely be possible to mitigate climate change. Green entrepreneurship promotes greater use of renewable energy. This book focuses on green entrepreneurship whose main target is both recycling waste products and the use of renewable energy sources. Green entrepreneurship has widely been adopted throughout the world. The world has acquired a number of green initiatives to mitigate climate change. In spite of its popularity, no meaningful researches have so far been conducted in this vital area in Bangladesh. For this book, the author focuses on some micro level green entrepreneurs as it is one of the vital sectors of the economy. If we can turn entrepreneurship into green entrepreneurship, it will be beneficial for both economic growth and preserving our environment. By overcoming the barriers and obstacles, and by enhancing the opportunities, if we can switch to 'Go Green', Bangladesh will be a better place for green business. It is expected that this book will open up a new dialogue between the government, media and development organizations which may introduce desirable and comprehensive policies that will support green entrepreneurship in flourishing in Bangladesh. Gupta and Barua (2018) argued that green entrepreneurship will complement the government's existing policy initiatives to mitigate climate change. The current initiatives are broad, including green financing schemes, national waste management and recycling programs and the promotion of green skills and green energy. The three million households using solar energy in Bangladesh are just one example of how serious the government is about greening the country. This book explores the challenges and possibilities of adopting green entrepreneurship in the context of climate change and sustainable development in Bangladesh.

Conceptual Framework

The conceptual framework consists of three independent factors such as green finance, government initiatives, attitudes and perception of green entrepreneurs. The green finance includes Infrastructure Development Company LTD (IDCOL), bank and other financing facilities. Financing capability is the prerequisite for any entrepreneurial development. Growth of green entrepreneurship depends on the availability of green financing. The second important factor is government support. The empathetic support from the government through their policies and activities is extremely essential. The third important factor is the attitudes, awareness, and perception of green and non-green entrepreneurs towards green entrepreneurship. The author argues that the challenges and possibilities of adopting a green model in businesses depends on these factors.

The conceptual framework consists of three important factors that impact on green entrepreneurship. The factors include green finance (Bank, IDCOL), the role of government and attitudes, and the awareness and actions of entrepreneurs regarding green entrepreneurship. These three factors have influence on green entrepreneurship. Bangladesh Climate Change Strategy and Action Plan (BCCSAP) recognized six thematic areas that demand proactive action from the Bangladesh Government so that the country can easily deal with the challenges of climate change. The thematic areas are: i) food security, ii) social protection and health; iii) comprehensive disaster management; iv) infrastructure; v) mitigation and low carbon development; vi) research and knowledge management and vii) capacity building and institutional strengthening. This is further helped by the nations that have extended their hands to support Bangladesh to promote its resilience against the adverse impacts of climate change. Although Bangladesh has not yet introduced any comprehensive national green finance strategy, Bangladesh Bank has recently adopted some strategies to promote green financing. Bangladesh Bank first issued Green Banking Policy Guidelines in 2011, which instructed commercial banks to adopt a comprehensive green banking policy in three phases.

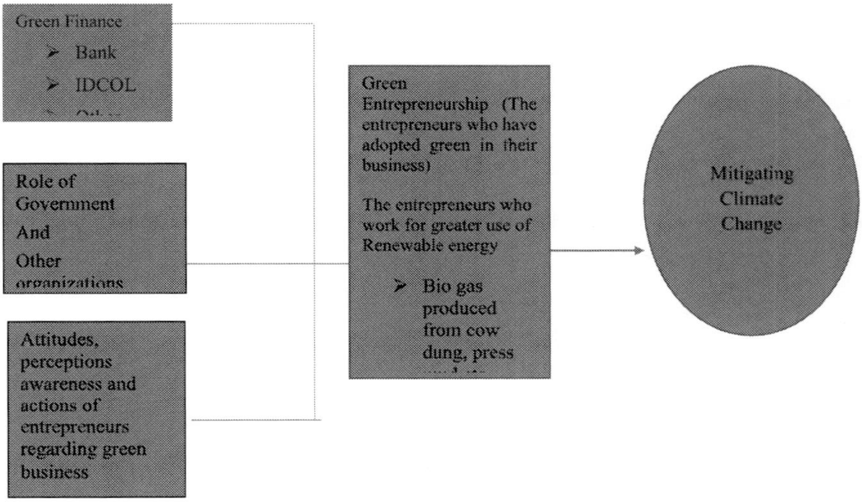

Source: The Author.

Figure 6. Conceptual Framework.

IDCOL has already adopted schemes for financing energy efficient equipment, green building and the manufacturing of energy efficient appliances. IDCOL also finances environment friendly auto brick projects. Financing energy efficient projects is a recent initiative of IDCOL (Green Climate Fund, 2017). The following sectors are eligible for IDCOL's financing:

- Solar home system program
- Domestic bio gas program
- Bio-gas based power projects
- Other Renewable energy projects (i.e., solar diesel hybrid solution for telecom BTSs, solar powered transportation, rooftop solar system, solar cold storage, battery charging stations, community biogas projects etc.)

In this book, the author conceptualizes green entrepreneurship as a process of adopting green practices and strategies in business. In other words, the entrepreneurs work for the greater use of renewable energy and sustainable practices. Although some studies have already been undertaken solely on climate change mitigation, there is hardly any studies conducted on climate change mitigation in relation to green entrepreneurship. Therefore, the book seeks to identify the gap in the current body of knowledge. This chapter illustrates how these concepts are operationalized, what activities could be included in the 'green' part of entrepreneurship, and what the core characteristics of green entrepreneurs are. The next chapter will shed light and explain sustainable development as a goal while further defining green entrepreneurship as a process for achieving this goal.

Chapter 3

Getting to Know About Sustainable Development

Bay of Bengal's funnel shape always keeps Bangladesh in a vulnerable situation. Therefore, natural disasters like flood, drought, cyclone are very common incidents in Bangladesh. Due to the frequent visit of natural disasters, government of Bangladesh has taken many measures in order to mitigate adverse impact of climate change. During any sorts of flood and cyclone, concerned authorities claim their readiness for ensuring disaster risk reduction. However, it is always very difficult to have confirmation about the consciousness of the local people and their attitude for moving to the cyclone shelter. Further, we have to consider the prospective damages due to weak embankment structure of the coastal zone (Huq, 2009). This chapter highlights the background of sustainable development which is the ultimate target to be achieved by developing countries like Bangladesh.

Sustainable development can be defined as development that meets the needs of the present without compromising the ability of future generations to meet their own needs (Bebbington, 2001). Therefore, it focuses on both present and future. Various organizations and institutions contributed in creation of the concept of sustainable development. The most significant contribution was done by United Nations. The United Nations is an organization of sovereign states, which voluntarily join the UN to work for world peace (Bebbington and Unerman, 2018). It formally came into being on 24th October 1945. At that time, it had 51 countries as members. Currently, 193 countries are UN members; the most recent to join was South Sudan in July 2011. United Nations consists of six main organs such as the General Assembly, the Security Council, the Trusteeship Council, the Economic and Social Council, the International Court of Justice, and the Secretariat (Groom, 2001).

The core objectives of United Nations include keep peace throughout the world, develop friendly relations among nations, help nations work together to improve the lives of poor people, to conquer hunger, disease and illiteracy, and to encourage respect for each other's rights and freedoms and work as a centre for harmonizing the actions of nations to achieve these goals (Ziring, et al. 2005). Since its establishment, United Nations organized a number of conferences, workshops and seminars aiming to achieve Millennium Development Goals (MDG) and the goals to of sustainable development. United Nations first set eight development goals for developing countries which should have been achieved by 2015. The goals include eradicate extreme poverty and hunger, achieve universal primary education, promote gender equality and empower women, reduce child mortality, improve maternal health, combat HIV/AIDS, malaria and other diseases, ensure environmental sustainability, develop a global partnership for development (Sachs, 2012). The following table shows the Millennium Development Goals and their related targets call for achieving the following outcomes by 2015.

Henry and Vollan (2014), Peng (2017) Caiado et al. (2018) cited in Shi et al. (2019) identified some limitations of MDGs. There were many subprojects that were unmet despite of universal primary education was achieved and the goal of providing better drinking water was achieved well ahead of time. Moreover, in some developing countries the goals of reduction of child mortality, maternal mortality by two thirds were not accomplished. And also, some developing countries could not achieve gender equality in their primary education and the goals of reducing the number of people affected by hunger. Then MDG updated and extended with special focus on environment that have to be accomplished for 2030.

The core environmental areas that are included in SDGs are sustainable food security, sustainable water security, universal clean energy, healthy and productive ecosystems, and governance for sustainable societies. The driving principles remain: reducing poverty and hunger, improving health and well-being and creating sustainable production and consumption patterns. A goal of improving lives and livelihoods, for example, would promote sustainable access to food, water and energy while protecting biodiversity and ecosystem services.

The term 'sustainable development' was first used in 1980 in worlds conservation strategy (Klarin, 2018). It considers the present without compromising the future generations to meet their needs. The Sustainable Development Goals were launched by United Nations member States in 2015. It was also known as global goals for ensuring sustainable development. It aimed to conclude poverty, protect the planet so that people can enjoy living in the earth. It has integrated 17 goals that aims to achieve by 2030. These goals are so interconnected that action in one area will surely affect in others. Sustainable Development Goals makes a good balance between social, economic and environmental sustainability. These goals are designed in such a way that brings zero poverty, hunger, AIDs and discrimination against women. According to Sustainable Development Solutions Network (2015: 6)

> "Sustainable Development Goals form a cohesive and integrated package of global aspirations the world commits to achieving by 2030. Building on the accomplishments of their predecessors the MDGs, the SDGs address the most pressing global challenges of our time, calling upon collaborative partnerships across and between countries to balance the three dimensions of sustainable development: economic growth, environmental sustainability and social inclusion."

Table 7. Millennium Development Goals, Targets and indicators

Millennium Development Goals	Related targets	Indicators
Goal 1	Decreasing by half the proportion of people living in extreme poverty and suffering from hunger.	Target 1: Proportion of population below $1 per day, Poverty gap ratio [incidence x depth of poverty], Share of poorest quintile in national consumption Target 2: Prevalence of underweight children (under five years of age), Proportion of population below minimum level of dietary energy consumption
Goal 2	Achieving universal primary education.	Target 3: Net enrollment ratio in primary education, Proportion of pupils starting grade 1 who reach grade 5, Literacy rate of 15–24-year old.
Goal 3	Eliminating gender disparity at all levels of education	Target 4: Ratio of girls to boys in primary, secondary and tertiary education, Ratio of literate females to males of 15–24-year-old, share of women in wage employment in the non-agricultural sector, Proportion of seats held by women in national parliament
Goal 4	Reducing the under-five mortality rate by two-thirds.	Target 5: Under-five mortality rate, Infant mortality rate, proportion of 1-year old children immunized against measles
Goal 5	Reducing the maternal mortality ratios by three-quarters.	Target 6: Maternal mortality rate, proportion of births attended by skilled health personnel
Goal 6	Halting and beginning to reverse the spread of HIV/AIDS, malaria, and other diseases.	Target 7: HIV prevalence among 15–24-year-old pregnant women, Contraceptive prevalence rate, number of children orphaned by HIV/AIDS Target 8: Prevalence and death rates associated with malaria, Proportion of population in malaria risk areas using effective malaria prevention and treatment measures, Prevalence and death rates associated with tuberculosis, Proportion of TB cases detected and cured under DOTS (Directly Observed Treatment Short Course)

Millennium Development Goals	**Related targets**	**Indicators**
Goal 7	Ensuring improved environmental sustainability	Target 9: Proportion of land area covered by forest, Land area protected to maintain biological diversity, GDP per unit of energy use (as proxy for energy efficiency) , Carbon dioxide emissions (per capita) Target 10: Proportion of population with sustainable access to an improved water Target 11: Proportion of people with access to improved sanitation, Proportion of people with access to secure tenure
Goal 8	Develop a Global Partnership for Development	Target 12: Official Development Assistance , Target 13: Market Access , Target 14: Debt Sustainability Target 15: Deal comprehensively with the debt problems of developing countries through national and international measures in order to make debt sustainable in the long term Target 16: In cooperation with developing countries, develop and implement strategies for decent and productive work for youth Target 17: In cooperation with pharmaceutical companies, provide access to affordable essential drugs in developing countries, Target 18: In cooperation with the private sector, make available the benefits of new technologies, especially information and communications

Source: Adapted from World Bank, 2002: 16–17, Soubbotina, (2004 :204).

Table 8. Seventeen Sustainable Development Goals Some provisional targets for 2030

Goals	Related Target	Indicators
Goal 1	End poverty in all its forms everywhere	Implement nationally appropriate social protection systems and measures for all,
Goal 2	End hunger, achieve food security and improved nutrition and promote sustainable agriculture	including floors, and by 2030 achieve substantial coverage of the poor and the vulnerable
Goal 3	Ensure healthy lives and promote wellbeing for all at all ages	Proportion of population covered by social protection floors/systems, by sex, distinguishing children, unemployed persons, older persons, persons with disabilities, pregnant women, newborns, work-injury victims and the poor and the vulnerable
Goal 4	Ensure inclusive and equitable quality education and promote lifelong learning opportunities for all	By 2030, ensure that all girls and boys have access to quality early childhood development, care and pre-primary education so that they are ready for primary
Goal 5	Achieve gender equality and empower all women and girls	education. Participation rate in organized learning (one year before the official primary entry age), by sex
Goal 6	Ensure availability and sustainable management of water and sanitation for all	By 2030, achieve universal and equitable access to safe and affordable drinking water for all 6.1.1 Proportion of population using safely managed drinking water
Goal 7	Ensure access to affordable reliable, sustainable and modern energy for all	services By 2030, expand international cooperation and capacity building support to
Goal 8	Promote sustained, inclusive and sustainable economic growth, full and productive employment and decent work for all	developing countries in water- and sanitation related activities and programs, including water harvesting, desalination, water efficiency, wastewater treatment, recycling and reuse technologies
Goal 9	Build resilient infrastructure, promote inclusive and sustainable industrialization and foster innovation	Amount of water- and sanitation related official development assistance that is part of a government Protect labor rights and promote safe and secure working environments for all
Goal 10	Reduce inequality within and among countries	workers, including migrant workers, in particular women migrants, and those in precarious employment

Goals	Related Target	Indicators
Goal 11	Make cities and human settlements inclusive, safe, resilient and sustainable	Increase in national compliance of labour rights (freedom of association and collective bargaining) based on International Labour Organization (ILO) textual sources and national legislation, by sex and migrant status Proportion of members and voting Proportion of members and voting rights of developing countries in international organizations
Goal 12	Ensure sustainable consumption and production patterns	
Goal 13	Take urgent action to combat climate change and its impacts	
Goal 14	Conserve and sustainably use the oceans, seas and marine resources for sustainable development	
Goal 15	Protect, restore and promote sustainable use of terrestrial ecosystems, sustainably manage forests, combat desertification, and halt and reverse land degradation and halt biodiversity loss	
Goal 16	Promote peaceful and inclusive societies for sustainable development, provide access to justice for all and build effective, accountable and inclusive institutions at all levels	Ensure public access to information and protect fundamental freedoms, in accordance with national legislation and international agreements Number of countries that adopt and implement constitutional, statutory and/or policy guarantees for RPD DOMAIN public access to information
Goal 17	Strengthen the means of implementation and revitalize the Global Partnership for Sustainable Development	Strengthen the means of implementation and revitalize the Global Partnership for Sustainable Development

Source: Adapted from United Nations (2015).

A set of six sustainable development goals (SDGs) follow from combining the Millennium Development Goals (MDGs) with conditions necessary to assure the stability of earth's systems.

New Paradigm

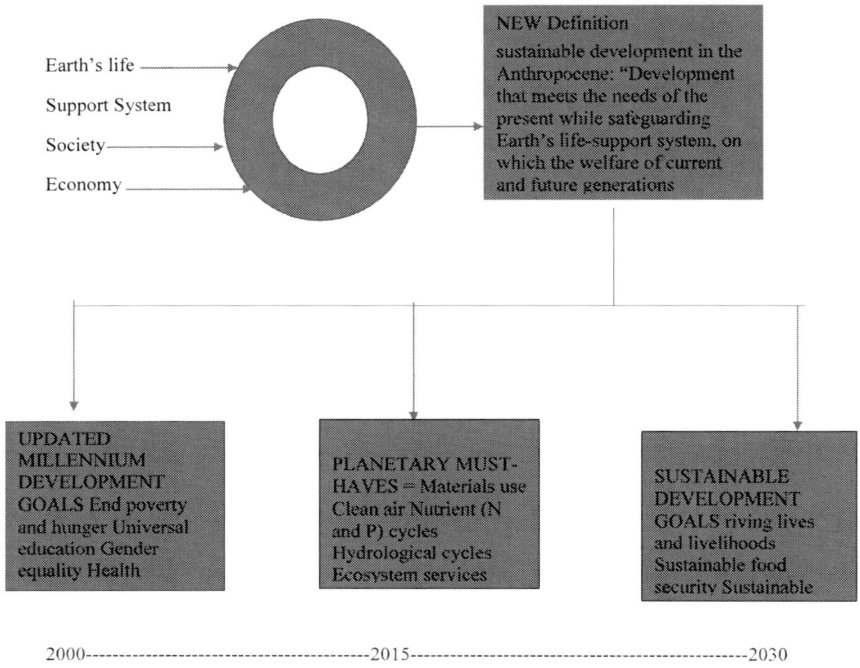

Source: Adapted from Nature (2003:3).

Figure 7. A Unified Framework of MDGs and SDGs.

Rudra and Kurian (2018), Muff et al. (2017) stated that United Nations adopted Sustainable Development Goals in 2015 based on the experiences they achieved from MDGs. SDGs includes six key elements dignity, human beings, planet, prosperity, justice and partnership. SDGs can be divided into four aspects such as i) economy related issues were described in goals eight, nine, ten and twelve. ii) society related issues were described in goals one, three, four, five, eleven and sixteen. iii) environment related goals were included goals two, six, seven, thirteen, fourteen and fifteen. iv) governance issues were identified in goal seventeen. SDGs consists of seventeen goals and one hundred and sixty-nine sub goals for all regions including developed and developing countries that have be achieved by 2030.

The outcome document for the SDG Agenda synthesizes the breadth of these issues by declaring that the SDG framework will stimulate action on five key themes: people, planet, prosperity, peace, and partnerships, which are described briefly below:

Table 9. Key Themes of SDGs

Key Themes	5 Ps
People	"We are determined to end poverty and hunger, in all their forms and dimensions, and to ensure that all human beings can fulfill their potential in dignity and equality and in a healthy environment."[1.6]
Planet	"We are determined to protect the planet from degradation, including through sustainable consumption and production, sustainably managing its natural resources and taking urgent action on climate change, so that it can support the needs of the present and future generations."[1.10]
Prosperity	We are determined to ensure that all human beings can enjoy prosperous and fulfilling lives and that economic, social and technological progress occurs in harmony with nature."[1.12]
Peace	We are determined to foster peaceful, just and inclusive societies, which are free from fear and violence. There can be no sustainable development without peace and no peace without sustainable development."[1.14]
Partnerships	We are determined to mobilize the means required to implement this Agenda through a revitalised Global Partnership for Sustainable Development, based on a spirit of strengthened global solidarity, focused in particular on the needs of the poorest and most vulnerable and with the participation of all countries, all stakeholders and all people."[1.16]

Source: Aadapted from sustainable development solutions network (2015:8).

Lele (2001) argued that sustainable development has been emerged as a latest development slogan over last few years. Various governmental as well as nongovernmental organizations have admitted sustainable development as latest paradigm of development. Klarin (2012) published a rigorous research in the concept of sustainable development: from its beginning to the contemporary issues. Sharpley (2000 cited in Klarin, 2012) argued that development and sustainability have counterproductive effects on the other hand Lele, 1991 cited in Klarin (2012) stated that there is no contradiction between sustainability and development. In the same article, Klarin (2012) argued that there is no development without sustainability or sustainability without development. Hopwood et al. (2005) wrote an article titled sustainable development: mapping different approaches. They stated in the article that sustainable development intends to combine environmental issues with socio economic issues. The article addresses the major challenges of humanity for now and future and focuses on sustainable livelihood and wellbeing. The concept of sustainable development comes from growing awareness regarding various issues including huge environmental problems, socio economic issues relating with poverty and inequality and it brings healthy future for humanity.

The main United Nations conference was held in Rio de Janeiro in June, 1992. It was recognized as The United Nations Conference on Environment and Development (UNCED), also known as the Rio Earth Summit. The conference was held in Rio de Janeiro 20 years after the historical UN conference on the human environment in Stockholm. The Stockholm conference was held in Sweden in June 5-16, 1972. It is also known as the United Nations Conference on the Human Environment. Sweden was the first country who first recommended the United Nations Economic and Social Council (ECOSOC) to arrange an International UN conference regarding human interactions with the environment. The UN conference on Human Development inspired many countries around the world to monitor environmental issues and to launch environmental agencies for doing it. One of the major outcomes of the conference was the establishment of United Nations Environment Program (UNEP). The UNEP assisted developing countries for implementing environmentally sound policies and practices. Despite of UNEP's establishment, failure to achieve most of its action programs have led to other conferences. Second Environmental conference was held in Nairobi (Kenya) in 1982.

Rio+10 was held in Johannesburg, South Africa. It is also known as the world summit on sustainable development which took place from 26th August to 4th September 2002. This conference was arranged after 10 years of the first earth summit in Rio de Janeiro in order to develop consensus on sustainable development by the United Nations. It was a sort of evaluation and follow-up on the implementation of the Rio-92 conference agreements.

The world summit on sustainable development was also known as earth summit ii that generated conventions on climate change, biodiversity and forestry. The summit also focused on combining economic growth with ecological accountability. The earth summit suggested a list of development practices named 'Agenda 21' which is a non-binding action plan of the United Nations in order to achieve global sustainable development. Agenda 21 is a roadmap for achieving sustainable development adopted in the conference held at Rio de Janeiro Rio+10 emphasized on integration of economic, environmental and social decision making.

Table 10. Overview of the various activities related to the concept of sustainable development

Year	Activities	Brief Description
1969	UN published the report Man and His Environment or U Thant Report.	Activities focused to avoid global environmental degradation. More than 2,000 scientists were involved in creation of this report.
1972	First UN and UNEP world Conference on the Human Environment, Stockholm, Sweden.	Under the slogan Only One Earth, a declaration and action plan for environmental conservation was published.
1975	UNESCO conference on education about the environment, Belgrade, Yugoslavia.	Setting up a global environment educational framework, a statement known as the Belgrade Charter.
1979	The First World Climate Conference, Geneva, Switzerland.	Focused on the creation of the climate change research and program monitoring.
1981	The first UN Conference on Least Developed Countries, Paris, France.	A report with guidelines and measures for helping the underdeveloped countries.
1984	Establishment of United Nations World Commission on Environment and Development (WCED).	The task of the Commission is the cooperation between developed and developing countries and the adoption of global development plans on environmental conservation.

Year	Activities	Brief Description
1987	WCED report Our Common Future or Brundtland report was published.	A report with the fundamental principles of the concept of sustainable development.
1987	Montreal Protocol was published.	Contains results of the researches on harmful effects on the ozone layer.
1990	The Second World Climate Conference, Geneva, Switzerland.	Further development of the climate change research and monitoring program and the creation of global Climate Change Monitoring System.
1992	United Nations Conference on Environment and Development (Earth Summit or Rio Conference), Rio de Janeiro, Brazil.	In the Rio Declaration and Agenda 21 Action Plan principles of sustainable development were established and the framework for the future tasks as well.
1997	Kyoto Climate Change Conference, Kyoto, Japan	The Kyoto Protocol was signed between countries to reduce CO2 and other greenhouse gas emissions, with commencement in 2005.
2000	UN published Millennium declaration.	Declaration containing eight Millennium Development Goals (MDGs) set by 2015.
2002	The World Summit on Sustainable Development, Johannesburg, South Africa.	Report with the results achieved during the time from the Rio Conference, which reaffirmed the previous obligations and set the guidelines for implementation of the concept in the future.
2009	The Third World Climate Conference, Geneva, Switzerland.	Further development of the global Climate Change Monitoring System with the aim of timely anticipation of possible disasters.
2009	World Congress Summit G20, Pittsburgh, USA.	G20 member states made an agreement on a moderate and sustainable economy.
2012	UN conference Rio +20, Rio de Janeiro, Brasil.	Twenty years from the Rio conference report. The future we want renewed the commitment to the goals of sustainable development and encouraged issues of the global green economy.
2015	UN Sustainable Development Summit 2015, New York, SAD.	The UN 2030 Agenda for Sustainable Development was published, setting up 17 Millennium Development Goals which should be achieved by 2030.
2015	UN conference on climate change COP21Paris Climate change Conference, Paris, France.	Agreement on the reduction of greenhouse gases in order to reduce and limit global warming

Source: Adapted from IISD, 2015; SDKP, 2015; UN, 2015ab; UNEP, 2015ab; UNDP, 2015c; WMO, 2015; UNFCCC, 2016. Cited in Klarin 2018: 72.

Table 11. The structure of Agenda 21

Section 1: Social and Economic Dimensions	Eight Chapters: Covering international co-operation, combating poverty, consumption patterns, heath, population, settlement and integrated environment and development decision making
Section 2: Conservation and management of resources for development	Fourteen chapter on environment: This includes atmosphere, oceans, fresh waters and water resources, land resource management, deforestation, desertification, mountain environments, sustainable agriculture, sustainable agriculture and rural development, conservation of biological diversity and biotechnology, toxic, hazardous, solid and radioactive wastes
Section 3: strengthening the role of major groups	Ten chapters discussing the role of women, young people and indigenous people in sustainable development, the role of Non-Government organizations, local authorities, trade unions, business, scientist and farmers
Section 4: Means of Implementation	Eight Chapters: how to pay for sustainable development, the need to transfer environmentally sound technology, science, the role of education, international capacity building, international legal instruments and information flows.

Source: Adapted from Robinson (1993 cited in Elliott, 2013).

Klarin (2018) argued that Agenda 21 is a global package that contains the purposes of sustainable development which is divided into 40 chapters. Agenda 21 provides a complete guideline for socio economic development in light of environmental conservation. The document highlights the necessity for international agreement between development and environment, UNDSD (1992 cited in Klarin, 2018). It emphasizes Government's responsibility to make and execute various plans, policies and programs. The documents also admit that for combining development and environment, it is highly essential to ensure other stakeholders' cooperation in this regard. The priority goal reduction of poverty along with preservation and protection of natural resources in poor countries.

Another International United Nations Conference on Sustainable Development named Rio +20 was organized in Rio de Janeiro in 2012. It was also known as earth summit on sustainable development that intended reconciliation the economic and environmental goals of the global community. World leaders along with participants from Governments, private sector, NGOs and other organizations joined in Rio+20 to discuss on various issues such as environment, sustainability, poverty and equity. Rio+20 was a 20-year follow-up of 1992 United Nations conference on Environment and Development. Rio+20 made a detailed plan to reevaluate economic development, ensure environmental protection and social equity. This conference focused on two core issues. One is a green economy in the context of sustainable development and the creation of institutional framework for achieving it. The conference identified seven priority areas which include decent jobs, energy, sustainable cities, food security, sustainable agriculture, water, oceans and disaster readiness. The book identifies the extent to what and how environment, climate change and green entrepreneurship are mutually interconnected for achieving sustainable development.

Chapter 4

THE GLOBAL AND NATIONAL CONTEXT

Climate change is the greatest global threat of twenty first century (Center for climate change and health, 2015). In this book, the author has explored various ways in which climate change and green entrepreneurship are connected. The author has engaged in fifteen- month process that included both field work and secondary research. The author has argued that human activity is the most significant cause of climate change and green entrepreneurship or green business could be one of the key approaches to lessen the causes of climate change. If green entrepreneurship is adopted, the greenhouse gas emissions to the atmosphere will be reduced. As mentioned earlier, sustainable development is a combination of society, economy and environment. it is obvious climate change (environment) and green entrepreneurship (economy) are powerfully interconnected to achieve sustainable development.

According to OECD (2011: 6)

> "Climate change refers to alterations of the earth's atmosphere leading to changes in the climate system, such as climate warming and more frequent and intense extreme weather events."

LEVEL 5: CONTEXT		
Institutional Physical infrastructure (e.g., urban design), Laws & regulations · Media & advertising	Social & Cultural, Physical Environment Prescriptive & descriptive norms Direct personal contact with influence agents Family, organizational and community structures	Physical Environment Climate demanding characteristics of residence (e.g., temperature)

LEVEL 4: INDIVIDUAL FACTORS	
Individual Factors Age, income, household size	Psychological Drivers Intrapersonal factors (e.g., needs, wants, goals) · Ideologies, beliefs, attitudes & worldviews about the environment Perceptions of prescriptive & descriptive norms

LEVEL 3: ECONOMIC CONSUMPTION	
Organization Behaviors Purchase goods & services Investment in equipment, production facilities & means of distribution of goods	Individual Behaviors Climate-driving characteristics & use of goods & services

LEVEL 2: ENVIRONMENTAL CONSUMPTION	
Organization Behaviors Climate-driving characteristics & use of equipment, production facilities & distribution of goods & services	Individual Behaviors Climate-driving characteristics & use of goods & services

LEVEL 1: GREEN HOUSE GAS EMISSIONS		
Carbon dioxide	Methane	Nitrous Oxide

LEVEL 0: CLIMATE CHANGE			
Temperature Change	Precipitation Change	Sea Level Rise	Extreme Events

Source: Adapted from a report of the American Psychological Association (2011: 34).

Figure 8. Conceptual model that shows context, individual factors, economic consumption, environmental consumption that leads to greenhouse gas emissions. The final consequence is climate change.

After a long debate there is now consensus that climate change takes place because of global average air and ocean temperatures, melting of ice and rising global sea levels. There are several consequences of climate change such as rainfall, snow and salinization, prolonged droughts etc. (Hensher, 2008). It is also widely acknowledged that human activities are also responsible for climate change. People are responsible for increased quantities of greenhouse gases emissions that has played significant role in climate change. People especially poor people are negatively affected by climate change (Hallegatte and Rozenberg, 2017). The author argues that the negative consequences of climate change make people's living condition more vulnerable.

There is another important dimension environment and climate change which is gender dimension. Environmental injustice has resulted many harmful health impacts on women. It is always women's responsibility to collect firewood and water for cooking. Women have to travel long distance and spend hours after hours for collecting firewood with empty stomach and bare foot. They come back with extreme hunger and cracked feet and a lot of injuries to their bodies. Sometimes they carry almost their body or little more than their body weight on their head that leads to high rates of uterine prolapse. As there is no alternative fuel for cooking women have to carry disproportionate burden on their head. In many countries including Somalia many displaced women have to travel hours after hours from their refugee camps for gathering firewood for cooking. Sometimes they have been raped or killed on their way (Nawaz, 2015).

Amartya (2004) stated that in Gujrat state of India women have to spend three or four hours for collecting firewood for cooking. With the rapid growth of deforestation women have to travel further and further and send longer hours than their mothers. An intergenerational study highlighted that women in Gujrat have to spend three four hours more than their mother did previously. Sometimes they have to carry 15 liters of water in each trip because fetching and carrying water is always women's responsibility. Aditi (2003 cited in Nawaz, 2015) stated that in the villages of desert district Banaskantha of India women have to spend 6 hours a day for collecting water from distant places. Banaskantha, which is in the country's western state of Gujarat, receives less than 7 inches of rainfall each year. The water table has dropped by 6.5 feet a year, as withdrawals exceed natural replenishment. Over 75% of the district's villages no longer have reliable, year-round sources of fresh water.

Environmental issues should be seen through gender perspectives because of the disproportionate burden that women have to bear. It is proved that there is some harsh environmental impact that only women experience through socially created gender division of labor imposed upon them (Preet et al. 2010). In various countries, women have to spend 12 or more hours a day inside and outside of home. In many parts of Africa and Asia women have to work 13 or more hours on an average than men do. Also, it's always women's responsibility to cook with biogas in indoor cooking spaces which creates indoor air pollution. Smoky toxic closed home environment is the most harmful of all chemical pollution. The author argues that environment and gender justice need to be integrated together (Batliwala, 1994, Nawaz, 2015).

In the context of environment and climate change, it is vital to do gender analysis because men and women are differently impacted by climate related hazards. Women are disproportionately impacted by climate related disasters. Therefore, men and women differently cope with climate change adaptation and mitigation strategies. In other words, since women have to follow socially created distinct roles and responsibilities that increases women's vulnerability and compels them to cope with climate change differently than men (Jonsson, 2011). Although globally more than half of women are involved directly in climate sensitive sectors such as agriculture and fisheries, both Kyoto protocol and UN framework Convention on Climate change did not even mention gender concerns (United Nations, 2019).

The author argues that Bangladesh is one of the most vulnerable countries to climate change in the world. Almost half of the total population are women who are involved in agriculture and economic activities. Globally women constitute more than half (51%) of total agricultural labor force but less than 20% women own land. Dankelman (2010) stated that women have neither right on nor they can give any decisions in agenda setting table despite of disproportionate burden of environment and climate change that women have to bear. Therefore, International Union for Conservation of Nature (IUCN) focuses on two things: one is to address the inequalities that prevails for women in the areas of biodiversity, climate change and disasters, energy, fisheries, agriculture, and other environment-related sectors. The other is open up opportunities for women as they are the agents of change and leaders in the environmental area (IUCN, 2020).

CONSEQUENCES OF CLIMATE CHANGE IN BANGLADESH

Bangladesh has been known as one of the largest deltas which is highly vulnerable to natural disaster such as flood, drought, cyclone etc. Almost 80% of its land area are vulnerable to flooding during rainy season. Moreover, there are some adverse effects of climate change such as such sea level rise, cyclones[1], salinity intrusion[2], heavy monsoon downpours, extreme temperature and drought[3] etc. Average temperature of Bangladesh is increasing during the monsoon season. A study shows that there has been significant increase in annual mean maximum temperature. The sea level has been risen gradually in last 30 years. Coastal destruction has been noticed. According to MoEF (2005) stated that Climate model scenarios estimate that mean annual temperatures will rise 1.0°C by 2030 and 2.4°C by 2100. Mean annual precipitation is expected to increase 5 per cent by 2030 and 10 per cent by 2100, and sea level is expected to rise 14 centimeters by 2030 and 88 centimeters by 2100. This is a major problem for the country, since much of its land is low-lying.

Warrick et al. (1993) reported that both Bangladesh and Netherlands are deltaic countries. However, Netherlands has scientific, financial and technical capacity to build sea walls to encounter climate change. They

[1] South and South-eastern Parts of the country were hit by Tropical Cyclones during the last few years. Approximately 130,000 people were killed in the cyclone of April 1990.

[2] Almost the whole Coastal Belt along the Bay of Bengal is experiencing Salinity problem.

[3] North and North-western regions of the country are suffering because of the Extreme Temperature problem.

have also taken various strategies to climate change which has to be incorporated with national and sectoral planning. On the other hand, Bangladesh is unable to take major initiatives like Netherlands because of its technical, scientific and financial incapacity but the country is taking some measures to reduce future emissions of greenhouse gas through the development of renewable energy. In rural areas, less than a quarter of its population are able to use electricity which means green energy needs to be supplied. Nonetheless, the country is yet to develop a compact plan of action to meet the energy needs and confront the challenges of climate change (Huq, 2001).

GREEN ENTREPRENEURSHIP: GLOBAL CONTEXT

World Economic Forum stated that entrepreneurs are considered as key drivers of economic progress and they are the real contributors to new job growth. Accordingly, Green entrepreneurs are called to have a leading role in the transition to economies that integrate the environmental capital as an asset for development and that are socially inclusive. Green entrepreneurs propose business models that are not only economically profitable but also create an environmental and social positive impact (Tharindu and Koggalage, 2020).

In a research- based paper entitled 'opportunities and Conditions for the Development of Green Entrepreneurship in the Polish Textile Sector,' Burzynska et al. (2018) argued that green or ecological entrepreneurship is closely associated with textile sector. Textile sector in Poland introduce various types of eco-innovations that prevent negative impact on environment. The core purpose of the paper is to analyze the prospects of green entrepreneurship from the perspectives of polish textile industry entrepreneurs. The findings of the research reveal that the entrepreneurs are interested in implementing all kinds of eco innovations. The following table depicts activities that are taken for promoting green entrepreneurship in some European countries.

Table 12. Activities for promoting the idea of green entrepreneurship in selected European countries

European Countries	Action
Bulgaria	Bulgaria Strategy 2020, Energy Efficiency and green economy, National Strategy for sustainable development of the Republic of Bulgaria, Program to support education for sustainable development
Romania	Presidency of 19th session of UN commission of sustainable development, green entrepreneurship program, National sustainable development strategy Romania 2020-2030, Romanian National Climate change strategy 2013-2020, Climate change and low carbon green growth program, National Strategy for competitiveness 2014-2020, National Strategy for Waste management 2014-2020, National Energy strategy 2015-2035, 3National Rural Development Program 2014-2020
Norway	Green and genius and energy for future programs, NOU 2015:8 reports
Greece	Green strategy, establishment of a separate Ministry of environment, energy and climate change
Latvia	National Development Plan 2020, Sustainable development strategy of Latvia 2030

Source: Adapted from Burzyńska, 2018.

Similarly, Uslu et al. (2015) conducted a green entrepreneurship research in Turkey. They argued that in the entrepreneurial ecosystem in Turkey, a number of organizations including non- governmental organizations supports this entrepreneurial approach. In Turkey, green entrepreneurship is a new and emerging sector, and, in many cases, environment is considered an important issue. This situation is recognized as an opportunity to improve the green entrepreneurial ecosystem. The authors reported various public policies such as tax breaks and tax exemptions facilitate green/eco entrepreneurial activities.

Some authors defined green entrepreneurship as sustainable entrepreneurship. Koe et al. (2015) conducted a research on factors influencing on sustainable entrepreneurship on SMEs in Malaysia. The authors argued that sustainable entrepreneurship is a possible way to handle environmental degradation. They stated that perceptual factors (perceived feasibility) and attitudinal factors (perceived desirability) are two core factors that positively and significantly increased the propensity of sustainable entrepreneurship than the normative factors. The authors suggested that for promoting sustainable entrepreneurship it is important to initiate sufficient sustainability capabilities, establish a positive or favorable sustainability attitude and create attractive sustainable entrepreneurial practice.

In other study, Čekanavičius*, L, Bazytė, R & Dičmonaitė, A (2014) compare and contrast between Lithuanuan and Irish companies. They argue considerable differences in the "green" attitudes and practices of Lithuanian and Irish companies, the former being more concerned with costs than benefits of becoming "green," and the latter demonstrating awareness that being 'green' opens new revenue-raising opportunities. Silajdzi c et al. (2015) conducted a research on green entrepreneurship in Bosnia and Herzegovina. This paper aims to explore a review of literature and statistical data on entrepreneurial environment. The authors conducted structured interviews with representatives of government, civil society, business sector and academia to understand the society's perception of green entrepreneurship. The findings of the research reveal that the entrepreneurs in these transitional countries are not ready to respond the challenges or to take any risks by investing in green business. The government and educational institutions do not play their role in supporting the development of green entrepreneurship.

Thang (2013) aims to advance an understanding of knowledge- based management practice in Vietnam. The author conducted a research on two Vietnamese agricultural companies. The authors argued that knowledge creation is highly influenced by human resource management, corporate culture. Leadership and team work. They further argued by providing illustrative examples of how knowledge-based management assists creativity of employees that eventually fulfil not only the profitability objective of the business but also added value in quality of life. O'Neill and Gibbs (2016) wrote a paper on green entrepreneurship to build a green economy. Evidence from UK. In this paper, the authors argued that the research on green entrepreneurship is so far been looked at separately, overlooking the wider socio -economic context. Therefore, it is important to research on green entrepreneurship holistically. The authors claimed that in UK, a new form of entrepreneurship has been developed combined with new demand from consumers for more environment friendly products. However, the authors have found a number of anxieties between policy intensions and business experience on the ground.

Banerjee and Dutta (2017) argued that green entrepreneurship must be adopted in order to minimize negative impacts of climate change on the society and biodiversity. Muo and Azeez (2019) wrote an article on green entrepreneurship: literature review and agenda for future research. Their purpose is to review, explore and integrate current analyses on green entrepreneurship and its connection with environment and sustainable development. For this purpose, the authors adopted integrative review of literature methodology. The findings of the article identified new trends in green entrepreneurship and identified the need to clarify some relevant concepts, such as: industry life cycle, entrepreneurship knowledge sharing, institutional framework, entrepreneurship financing, green entrepreneurship decision-making process among others.

Alwakid and Urbano (2020) conducted an in-depth study on green entrepreneurship in Saudi Arabia. The authors argued that recent discourse on green entrepreneurship is very much attached with cultural changes. While some countries are unable to adopt environmental advantage, some other countries are transitioning and progressing toward green entrepreneurship. They further stated that Saudi Arabia had taken some initiatives toward ecologically friendly society. By analyzing cultural characteristics with green entrepreneurship in Saudi Arabia. The authors pointed out that environmental actions, environmental consciousness, and temporal orientation, increase the level of green entrepreneurial activity are seen across the cities in Saudi Arabia. They also concluded the paper with the implications for policy and practice related to environmentally-friendly productive activities.

In Austria, Sustainability Balanced Scorecard (SBS) has recently been used in small medium enterprise. The paper highlights when developing the scorecard, it is essential to consider specific SME characteristics into account. The author Falle et al. (2016) also stated that it is important to apply methods that permit for adaptation to the existing conditions of SMEs such as strategic management, organizational structure and resources. In Austrian SMEs, sustainability management is ensured through sustainability Balanced Scorecard (SBS).

Hansen and Schaltegger (2018) conducted a study in one of the four largest German clothing retailers with a focus on organic cotton. The organic cotton practices have been pioneered by Hess Natur in 1970. The author stated that in the beginning the organic cotton concept was used in only one markets called niche but gradually it has been spread among mainstream retailers. The authors argued that their study on organic cotton will help entrepreneurs to understand their role in transformation of sustainability issue and how the transformation may affect them. The authors concluded in order to get success in the market, it is important to understand the transformation challenge and coping mechanisms of introducing organic practices.

Anabaraonye et al. (2019) conducted an in-depth study on green entrepreneurial opportunities in climate change adaptation and mitigation for sustainable development in Nigeria. They explored all possible entrepreneurial opportunities in climate change adaptation. The authors argued that climate change is a global phenomenon which needs to be tackled urgently for ensuring sustainable development in Nigeria. It is essential to introduce climate change education in order to prepare Nigerian communities to effectively deal with the impact of climate change. The inclusion of climate change poems in climate change education promotes sustainable development in Nigeria. This will further promote green entrepreneurial opportunities. The use of climate change poems and blogs in climate change education for sustainable development in Nigeria are also identified as green entrepreneurial opportunities and highly encouraged. Climate change education also trains people to mitigate and adapt climate change effectively. The authors suggested green entrepreneurial opportunities for maximizing economic opportunities in order to mitigate and adapt climate change. Green entrepreneurship includes adopting more efficient renewable energy sources, strategic tree planting activities, radical awareness approach of information dissemination and adopting good waste management and recycling strategies. After conducting both primary and secondary research, the authors stated that in Nigeria many communities, institutions and individuals are needed to promote green entrepreneurship. Some green institutions are badly needed who would provide green jobs and green skills to unemployed.

Green Entrepreneurship in Asian Context

Sharma and Kushwaha (2015) argues that in India, consumers are now health and environment conscious. Perhaps this is the reason, that the concept of green product and green marketing are taking place slowly and steadily in the market. It can be also as an opportunity for those companies, who believe in innovation and redesigning their products and services in ecological manner. It is a perfect time for those entrepreneurs who want to come and be a part of this emerging green market. The authors stated that there is a bright opportunity for the green entrepreneurs as the green market is expanding and consumer behavior is favorable towards green market. In their article, Farinelli et al. (2011) argued that a Green Economy needs to be driven by entrepreneurs that respond to policy incentives through innovation in management and technology.

In their paper, Sharma and Kushwaha (2015) argued that in India consumers taste and preferences have been shifted a lot in last two decades. They stated that in recent market scenario consumers are health, environment conscious. For this reason, the concept of green product and green marketing are taking place in the existing market. The authors wanted to explore the opportunities for green entrepreneurs in the transforming scenario of market shift. On the basis of available literature, the findings of the study revealed that there is a bright prospect of green entrepreneurs in emerging green market in India which is one of the best places in the world for investment from this point of view.

Mathur and Tandon (2016) explored the potential opportunities and challenges faced by green entrepreneurs in India. The paper offers a sustainable solution to prevent further ecological degradation through green entrepreneurship. In this research, the authors chose 130 students through random sampling method to reveal the prospects and key barriers of green entrepreneurship in India. The study reveals that the millennials possess high level of environmental awareness. The opportunities include emergence of various green sectors in Indian economy. On the contrary, the challenges include limited knowledge of green technology and high investment costs, absence of funds to undertake risk in the sector along with inability to understand potential benefit accrued from green businesses.

Halder (2018) wrote a research paper entitled Green entrepreneurship in the renewable energy sector – a case study of Gujarat. The article reveals some critical factors that shape the green entrepreneurship in solar energy sector in Gujrat. In his paper, the author provides valuable insights on barriers and motivations and policy suggestions to foster an environment of green entrepreneurship in solar energy sector in Gujrat, India. The author argued that there is huge opportunity of green entrepreneurship in India but at the same time green entrepreneurs will serve as drivers of change who can transform people's mindset towards green market.

Sambrani, (2016) conducted a study on Horticulture (vegetable plants nursery) entrepreneurship in India. The author argued that rural development is must for all out development in India because 70% of its population live in rural areas. They also argued that rural development is linked to rural entrepreneurship because it offers a huge potential for employment. The author discussed about an entrepreneur who runs horticulture and helps the neighboring farmers. The author tried to understand Government's policies to support horticulture entrepreneurship. The author concluded that government of India provides subsidies under various programs to support horticulture entrepreneurship. Thus, India has progressed a lot in green entrepreneurship through various programs.

Gunawan and Fraser (2017) conducted a research on green and young entrepreneurs in Indonesia. The authors stated that since the unemployment rate is quite high in Indonesia, entrepreneurship is now being promoted by government as policy direction to promote economic and social equality. The article offers a relationship model between entrepreneurial characteristics, institutional environment and young entrepreneurs' innovativeness. The findings of the study reveal that there was not enough support from government both financially and institutionally. The authors argued that transport cost, lack of awareness of green products, process and services are major obstacles of green entrepreneurship in Indonesia.

From the above discussion, it is evident that a number of green initiatives have been adopted in many European countries. And the government of these European countries have realized how important it is to take green activities for the sake of future of these countries. Many Asian countries especially India has taken a number of green initiatives. The studies revealed that there is a bright prospect of green entrepreneurs in emerging green market in India. Like many other countries of the world, green entrepreneurship has also gained popularity in Bangladesh. Although some researches on green entrepreneurship have been conducted in European and other south Asian countries, there is hardly any meaningful research conducted in Bangladesh. Therefore, it is timely to conduct a comprehensive research on green entrepreneurship in relation to climate change mitigation.

It is also important to highlight that men and women are disproportionately affected by environment and climate change. Therefore, the author intends to link the dots of environment, climate change and women. Women carry a disproportionate burden of unpaid and underpaid work. There is sharp socially created gender division of labor that prevails and perpetuates throughout the world specially in developing and under developed countries. Globally 70% of unpaid and underpaid work are done solely by women. Women in Asia spend far more time than men for doing unpaid household work. Women have to give domestic labor regardless of their productive work and financial contribution to family (Antonopoulos, 2009).

CGC (2008) reported that in South East Asia women constitute more than half (51%) of the total agricultural labor force. Women spent 90% of the labor to cultivate rice, vegetables, fruit and grain for home consumption and sometimes for sale in rural areas. Women do not take safety measures when they use toxic chemicals and pesticides in the air, water and earth are responsible for a variety of women's health risks. In Gansu province in China state run fertilizer factory have led to an excessive number of still births and miscarriages. Similarly, in Sudan a positive correlation has been found between pesticides and perinatal mortality. Many women are at high risk while using pesticides because it leads to perinatal mortality (Batliwala, 1994 cited in Nawaz, 2015).

Moreover, women farmers are not protected when handling insecticide, pesticides and fertilizer which can lead to a very high chance of sterility. Women's incapability of producing children increases the chance of domestic violence. Sometimes husbands remarry due their wife's incapability of producing children. Likewise, in South Africa many women are trying to help their families through cultivation. But women have to depend hugely on chemicals and pesticides that leads to lot of gynecological disorders. With the rapid growth of deforestation and desertification, all these unprotected tasks still remain women's responsibility and this work has become harder and harder by day.

Climate Change Mitigation and Green Entrepreneurship

Climate change is one of the most crucial issues that the world is facing today. It is a global phenomenon that includes science, economics, society, politics, moral and ethics. Responding to climate change involves two major approaches; i. climate change mitigation ii. Climate change adaptation. Climate change mitigation refers to reduction of climate change through minimizing greenhouse gas emissions into atmosphere. The following figure shows how greenhouse gases warm our planet.

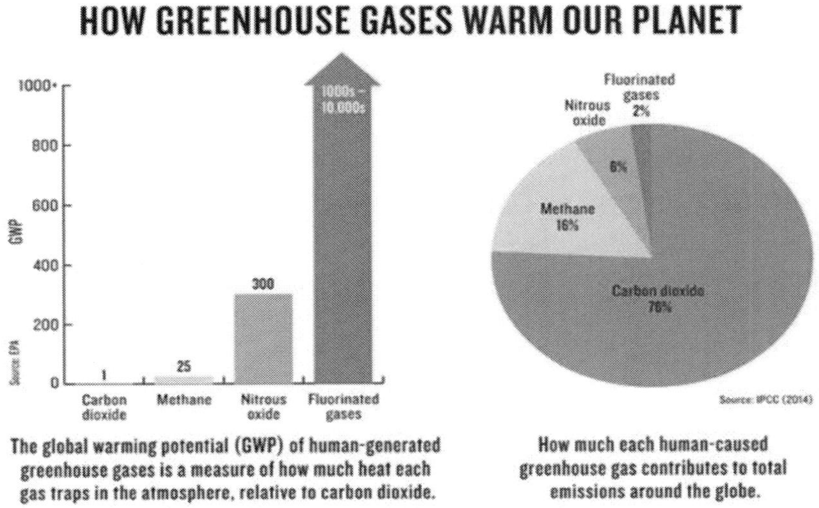

Source: IPCC (2014).

Figure 9. How does greenhouse gas warm our planet.

Greenhouse gas includes carbon dioxide, methane, nitrous oxide and fluorinated gases. The pie chart shows how much each human caused greenhouse gas contributes to total emissions around the globe. It shows carbon dioxide 76%, methane 16%, nitrous oxide 6% and fluorinated gases 2%. United Nations Intergovernmental Panel on Climate Change (2014) reported that the ultimate purpose of climate change mitigation is to prevent human from interfering with the climate system. Reduction of climate change involves diminishing the sources of harmful gases such as the burning of fossil fuels for electricity, heat or transport. In this book, I conceptualize green entrepreneurs to those who are both aware of environmental issues, ensures greater use of renewable energy and whose business venture is in the environmental marketplace. Such entrepreneurs pursue environmental-centered opportunities which show good profit prospects. I have chosen innovative green entrepreneurs who are trying to identify new opportunities to earn profit and protect environment in a sustainable way.

The government has adopted a number of policies where environment and development issues have been addressed. Important policy documents in this respect are the Environment Policy (1992), the Forest Policy (1994), the Fisheries Policy (1998), the Water Policy (1998), the New Agriculture Extension Policy (1995), The Energy Policy (1995). Besides these sectoral policies, the National Conservation Strategy (NCS) and especially the National Environment Management Action Plan, 1995 (NEMAP) have been formulated to provide action plans to respond to environmental issues and promote sustainable development. These are potentially important in guiding and influencing the activities of the concerned Ministries and other governmental agencies. The policies of Bangladesh Government have been developed mostly from a sectoral approach (ADB 2004 cited in Akter, 2009).

NAMA (Nationally Appropriate Mitigation Action)

Dion et al. (2012) stated that in developing countries NAMA has been taken to reduce greenhouse gas emissions to the atmosphere. NAMA indicates some policies, programs and projects that should have been adopted by developing countries in order to mitigate climate change and contribute to sustainable development. NAMA has become the key element in the whole mitigation framework of the United Nations Framework Convention on climate change (UNFCCC). Some developing countries have detailed NAMA while some others are in the process of developing detailed plans to implement these actions. Bangladesh has already developed Climate change strategy and Action Plan and made a significant progress in planning for adaptation. The country has also expressed her interest in undertaking mitigation strategies in order to achieve sustainable development goals. Therefore, Bangladesh has adopted NAMA which would be beneficial to reduce vulnerabilities of climate change (Dion et al. 2012).

Poor air quality is a major concern in urban centers and industrial areas of Bangladesh. However, because of larger emission sources (vehicles, brick kilns, road dust) and high impact possibilities (large population), Dhaka is of particular importance. Air quality data of other major cities (Chittagong, Khulna, Rajshahi) also show similar pattern. Biomass continues to play a critical role in the rural energy sector in Bangladesh. Like in many developing countries, biomass is extensively used in rural areas of Bangladesh for cooking; and fuel-wood, crop residue, cow dung, tree leaves and grass account for about 97% of total household energy use. Some 95 percent of rural households collect or purchase biomass for cooking, using mainly traditional clay stoves. Still less than 50% people in rural areas have access to electricity. In rural Bangladesh, some 70% of lighting energy is derived from kerosene (Asaduzzaman et al. 2010) through use of kerosene lamps (wick lamp and hurricane). Only limited data are available on indoor air quality in Bangladesh; almost all studies on indoor air quality focus on cooking using biomass fuel, without much reference to possible contributions from wick lamp/ hurricane.

WASTE MANAGEMENT

Second national communication to UNFCC stated that the major sources of methane emissions include domestic wastewater, livestock, paddy, poultry, and solid waste. In other words, wastewater is the most dominant source of methane emission. While wastewater constitute 39% of estimated total methane emission, solid waste account only 1% of the total emission. In Bangladesh it is quite common to burn harmful dry refuse such as plastics, metals and batteries. During winter season, it is fairly common to burn dry leaves which produces black carbon. These multiple types of wastes pollute environment and responsible for climate change.

Government of Bangladesh has separate policies for both climate change adaptation and climate change mitigation. Bangladesh Government has adopted Nationally Appropriate Mitigation Action (NAMA) in response to the United Nations Framework Convention on Climate Change

(UNFCCC). The GPRB also developed the Bangladesh Climate Change Strategy and Action Plan in 2008, National Adaptation Plan of Action in 2005 and National Communications in 2002. In addition, the Bangladesh Climate Trust Fund has been created, with domestic resources and the multi-donor Climate Resilient Trust Fund allowing the government to implement the Bangladesh Climate Change Strategy and Action Plan with the support from developed country parties. Bangladesh has shown leadership at UNFCCC negotiations and has been a driver of positions of the least developed country negotiating bloc. The country has expressed an interest in undertaking mitigation activities that contribute to the country's development goals and that entail adaptation and sustainable development (i.e., environment, economic and social) co-benefits. This interest and initiative in adapting to climate change, as well as Bangladesh's vulnerability to its effects, mean that the country could benefit significantly from developing and seeking funding for NAMAs, and is well positioned to do so.

While Bangladesh tends to focus on adaptation, given the country's vulnerability to climate change, mitigation also requires serious attention to ensure the country pursues less harmful gas emissions. Apart from agriculture, industry, transport, forestry and land use, Government gives priority to energy sector in the following areas: supply of energy/meet demand, making energy sector financially viable, increasing the efficiency of the sector and of energy use, enjoying energy security and pursuing sources of renewable energy (Dion, et al. 2012). There is little scope for additional power generation from gas and Bangladesh will increasingly be looking to its coal reserves to increase the supply of electricity (MOP, 2010). But there are also renewable energy options present in the country. Solar energy has real prospects in Bangladesh, and wind energy is feasible.

From NAMA report, it is clearly seen that renewable energy sector is given top priority to mitigate climate change. Eighteen NAMA opportunities were identified for Bangladesh: three for the agriculture sector, seven for the energy sector, two for the forestry sector, two for the industry sector, two for the transport sector and two for the waste sector. This report highlights both energy and waste management as a key strategy to mitigate climate change. Energy sector includes develop solar, wind and hydro renewable capacity, make cooking less carbon intensive and make use of waste to fuel technologies. Waste sector includes two issues i. improve systems of waste management ii. disposal methane capture and electricity production (MOEF, 2002, GRPB, 2008).

Table 13. Long listed NAMA opportunities for the energy sector

Measure	Source
Make infrastructure more efficient—less loss and waste	BOI (n.d.); MOEF (2002); GPRB (2008); MOP (2010)
Gas exploration and reservoir management	GPRB (2008)
Develop fossil fuel resources in as climate-friendly a way as possible—e.g., replace old power plants using improved and new technologies, clean coal technology	MOEF (2002); GPRB (2008); MOEF (2011)
Develop solar renewable energy capacity	BOI (n.d.); MOEF (2002); GPRB (2008); Asian Development Bank (ADB) (2011); MOEF (2011)
Develop hydro renewable energy capacity	GPRB (2002); GPRB (2008); ADB (2011)
Encourage energy conservation, introduce demand management measures	BOI (n.d.); MOEF (2002); MOP (2010)
Efficient building design, refrigeration, more energy-efficient lighting (e.g., Compact Florescent Lamps [CFLs]) and cooling	BOI (n.d.); MOEF (2002); GPRB (2008); MOEF (2011)
Make cooking less carbon-intensive: substitute fuels, have more efficient stoves, use solar cookers	BOI (n.d.); MOEF (2002); AWG-LCA presentation
Make use of waste-to-fuel technologies	BOI (n.d.); MOEF (2002)

Source: Dion, 2012.

After analyzing NAMA, this book sheds light on the policy gap in enhancing bio gas renewable energy capacity and producing electricity by using waste. It should have been introduced in NAMA. There is another major side of managing climate change which is climate change adaptation. While climate change mitigation indicates the efforts to diminish or stabilize Green House Gas emissions that involve forest management or energy conservation interventions, adaptation involves assisting people to cope with the negative consequences of climate change that involves targeted poverty reduction, asset building, early warning planning and disaster response planning. However, it is believed that there is a continuum between mitigation and adaptation, and therefore more integrated approaches are needed. In response to climate change, the Government of Bangladesh (GoB) and international agencies have placed high priority on enhancing the capacity of the country to mitigate and adapt to climate change. Therefore, the country has fostered two primary funding opportunities to deal with climate change. One is Bangladesh climate change trust fund (BCCTF) which has been created by GOB. The other is Bangladesh Climate Change Resilience Fund (BCCRF) which has been created by international development partners and monitored by the World Bank (WB).

Both BCCTF and BCCRF have contributed to the implementation of the Bangladesh Climate Change Strategy Action Plan (BCCSAP). BCCSAP is a national strategy and action plan for decreasing socio economic vulnerabilities to climate change. Not only GOB but also other Non-Government Organizations (NGOs) and Community Based Organizations (CBOs) have started addressing the immediate effect of climate change. However, all these organizations focus primarily on climate change adaptation thorough disaster risk management and climate change adaptation initiatives. This book argues that climate change mitigation also deserves high priority in respect of long-term solutions to climate change. For this reason, this book is timely as it is based on green entrepreneurship as an important strategy to mitigate climate change. This book is based on those entrepreneurs who are contributing in the field of renewable energy. The last chapter sheds light on the core findings and analysis which is followed by some key policy suggestions.

Chapter 5

LESSONS AND LOOKING FORWARD

Bangladesh is one of the most vulnerable countries in the world to the effects of climate change. In the last fifty years, the country has encountered a huge number of climate- related disasters including flood, drought and cyclone. These disasters have killed thousands and thousands of people destroyed many homes and livelihoods. The Asia Foundation (2012) reported that climate forecasts of Bangladesh always put us in additional uncertainty about the timeframe, scale and frequency of life-threatening catastrophes. The loss of biodiversity, a rising sea level, devastating cyclones, and frequent floods and droughts are linked to the rise in global temperatures.

Therefore, the book argues that it is essential to stimulate green entrepreneurship to mitigate the adverse impact of climate change in the long run. However, there are many challenges to promote green entrepreneurship. According to Hamid et al. (2013), key barriers of biogas technology dissemination include uncertainties about post-warranty services, high initial costs, scarcity of feedstock and the general reduction in the number of domestic animals, particularly cows. The book suggests that there are many challenges in promoting green entrepreneurship. The issues are discussed in the following with empirical data. Challenges have been divided into three broad categories: behavioral, financial and institutional.

Behavioral Challenge

Table 14. Knowledge about the term 'Green entrepreneur'

Knowledge	Number of green entrepreneurs	Percentage (%)	Number of non-green entrepreneurs	Percentage (%)
Good Knowledge	6	40%	1	10%
Moderate knowledge	3	20%	5	50%
Little knowledge	6	40%	2	20%
No Knowledge	0	0%	2	20%
Total	15	100%	10	100%

Source: Field Investigation.

This table illustrates that only 40% green entrepreneurs had good knowledge about it while only 20% had moderate knowledge and 40 % had little knowledge about it. It is quite surprising that entrepreneurs who are adopting green do not have adequate knowledge about it. On the other hand, one non -green entrepreneur had good knowledge about it. He spoke over two days regarding various aspects of green entrepreneurship and its prospects and challenges. The table demonstrates that 50% of non-green entrepreneurs had moderate knowledge whereas 20% had little knowledge and rest 20% had completely no knowledge about green. It means that green entrepreneurs do not have sufficient knowledge about the concept while some non-green entrepreneurs have adequate knowledge about it.

Table 15. Perspective regarding promoting green entrepreneurship

Opinion	No of green entrepreneurs	Percentage (%)	No of non-green entrepreneurs	Percentage (%)
Yes	8	53.33%	2	20%
No	7	46.67%	8	80%
Total	15	100%	10	100%

Source: Field investigation.

This table illustrates that only 53.33% green entrepreneurs said 'yes' when they are asked about the importance of promoting green entrepreneurship. On the other hand, 46.67% green entrepreneurs answered 'No' when they were asked whether they think it is essential to promote green entrepreneurship. It is surprising that those who are adopting green, majority of them still think that it is not important to promote green entrepreneurship. On the contrary, only 20% non- green entrepreneurs think that it is essential to facilitate green entrepreneurship while 80% of them believe that it is important to promote green entrepreneurship. Attitude of the entrepreneurs play an important role in eco entrepreneurship development because of its innovative nature.

Phan et al. (2002), Tonglet et al. (2004) and Schick et al. (2005) also found same findings in their studies. This study suggests that almost half of the respondents showed positive attitude towards eco entrepreneurship. Ajizen (1991 cited in Koea at al 2015) believes that when a person shows positive attitude towards certain behavior it is more likely that the person will perform the behavior.

While the author asked the participants whether they know about any government program/ policy that facilitate green entrepreneurship. The respondents replied in the following way:

Table 16. Knowledge about Government Program

Opinion	No of green entrepreneurs	Percentage (%)	No of non-green entrepreneurs	Percentage (%)
Yes	0	0%	0	0%
No	15	15%	10	100%
Total	15	100%	10	100%

Source: Field investigation.

This table demonstrates that none of the green entrepreneurs and non-green entrepreneurs had knowledge about government program/ policy on green entrepreneurship. It is surprising that when the author asked them whether they know about e any government policy or programs on green entrepreneurship, they instantly replied 'NO.' They also added that there is no government policy for facilitating green entrepreneurship outside Dhaka city. From their response, it seems that the government does not have adequate programs for promoting green entrepreneurship outside of the capital of Bangladesh.

Moreover, when the author asked the respondents whether they are familiar with the training programs on green entrepreneurship. They gave the following answers.

Table 17. Familiarity with training programs on green entrepreneurship

Opinion	No of green entrepreneurs	Percentage (%)	No of non-green entrepreneurs	Percentage (%)
Yes	0	0%	0	0%
No	15	100%	10	100%
Total	15	100%	10	100%

Source: Field Investigation.

This table demonstrates that none of the green entrepreneurs and non-green entrepreneurs is familiar with training programs on green entrepreneurship. It is surprising that when the author asked them whether they know about any training programs on green entrepreneurship, all of them replied 'NO.' All of them stated that it is very important to initiate various training programs on green entrepreneurship. Although they are interested in attending training programs, they have no opportunity available for them. It is evident from their response that there are no sufficient training programs for green entrepreneurs in Bangladesh. Besides, while the author asked the respondents regarding the potential factors would encourage them to implement a green (ecological) entrepreneurial approach in their enterprise. The author argues that entrepreneurs who decide to introduce innovations always expect benefits, but it is important to remember that motivation to develop innovations in enterprises should be based on the argument of long-term development rather than on one-off benefits. Entrepreneurs asked about incentives which convinced them to introduce the principles of green entrepreneurship. Eighty per cent of them indicated that they are inspired by successful entrepreneurs from other countries. Twenty per cent of the respondents emphasized that they would be encouraged by various training sessions or conferences which would increase the respondents' awareness of the costs and benefits of implementing eco-innovations. Therefore, from their responses, it is found that most of the green entrepreneurs adopted green business being inspired by other successful green entrepreneurs from other countries. They argued that if it is possible in other countries why not in Bangladesh.

Financial Challenge

The author asked the respondents whether they know about green baking policy and strategy framework formulated by Bangladesh Bank?

Table 18. Knowledge about green banking policy and strategy framework formulated by Bangladesh Bank

Opinion	No of green entrepreneurs	Percentage (%)	No of non-green entrepreneurs	Percentage (%)
Yes	0	0%	0	0%
No	15	100%	10	100%
Total	15	100%	10	100%

Source: Field Investigation.

The above-mentioned table demonstrates that none of the green entrepreneurs and non-green entrepreneurs know about green baking policy and strategy framework formulated by Bangladesh Bank. When I asked them whether they know about green banking policy and strategy framework formulated by Bangladesh Bank, all of them replied immediately that they do not know about green banking policy and strategy framework formulated by Bangladesh Bank. It is evident from the primary data that it is quite far away from its implementation. The author also asked the participants about their knowledge regarding environmental risk guidelines formulated by Bangladesh Bank, central bank of Bangladesh?

Table 19. Knowledge about environmental risk guidelines formulated by Bangladesh Bank

Opinion	No of green entrepreneurs	Percentage (%)	No of non-green entrepreneurs	Percentage (%)
Good Knowledge	2	13.33%	0	0%
Moderate knowledge	5	33.33%	0	0%
Little knowledge	5	33.33%	2	20%
No Knowledge	3	20%	8	80%
Total	15	100%	10	100%

Source: Field Investigation.

This table demonstrates that only 2 (13.33%) green entrepreneurs had good knowledge about environmental risk guidelines formulated by Bangladesh Bank while none of the green entrepreneurs had good knowledge about the environmental risk guidelines formulated by Bangladesh Bank. It is quite surprising that entrepreneurs who are adopting green do not have good knowledge about environmental risk guidelines. On the other hand, while only 5 (33.33%) had moderate knowledge regarding it, none of the non-green entrepreneurs had moderate knowledge about the issue. Again, 5 (33.33%) had little knowledge about environmental risk guideline and 3 (20%) completely had no knowledge regarding it. Nonetheless, only 2 (20%) and 8 (80%) non- green entrepreneurs had little and no knowledge respectively. Consequently, it is evident that very few entrepreneurs know about environmental risk guidelines formulated by Bangladesh Bank because it is not widely circulated. When the author asked the participants whether they have any knowledge about green/eco banking they replied in the following way.

Table 20. Knowledge about green/eco banking

Opinion	No of green entrepreneurs	Percentage (%)	No of non-green entrepreneurs	Percentage (%)
Yes	10	66.67%	5	50%
No	5	33.33%	5	50%
Total	15	100%	10	100%

Source: Field Investigation.

This table illustrates that 10 (66.67%) answered yes instantly when the author asked them whether they know about green/ eco banking whereas only 5 (50%) non-green entrepreneurs had knowledge about green banking. Five (33.33) % green entrepreneurs did not know about green banking nevertheless 5 (50%) non- green entrepreneurs had no knowledge about green banking. Then again when the author asked them whether they ever had green finance through green banking. All of them replied that they never acquired green finance through green banking. Moreover, while the author asked the respondents whether they know about green credit card facilities. None of them said that they know about green credit card.

Table 21. Knowledge about green credit card facilities

Opinion	No of green entrepreneurs	Percentage (%)	No of non-green entrepreneurs	Percentage (%)
Yes	0	0%	0	0%
No	15	100%	10	100%
Total	15	100%	10	100%

Source: Field Investigation.

This table illustrates that none (0%) of the green and non-green entrepreneurs answered yes instantly when I asked them whether they knew about green credit card facilities of bank. On the other hand, 15 (100%) green entrepreneurs and 10 (100%) non-green entrepreneurs answered that they had no knowledge about green credit card facilities. Moreover, when I talked with some bank officials they informed me that green credit cards were not available for entrepreneurs living outside of

Dhaka. Then the author asked whether the respondents are satisfied with IDCOL's (Infrastructural Development Company Limited) financial facility.

Table 22. Satisfaction with IDCOL's financial facility

Opinion	No of green entrepreneurs	Percentage (%)
Yes	0	0%
No	15	100%
Total	15	100%

Source: Field Investigation

This table illustrates that none (0%) of the green entrepreneurs is satisfied with financial facility provided by IDCOL. All of them (100%) reported that they are highly dissatisfied with IDCOL's policy for many reasons. Some of them stated that they went to IDCOL to access their green loan but they were informed that IDCOL's financing facility is only available for the entrepreneurs living in Dhaka city. Therefore, it evident is that green entrepreneurs living outside Dhaka are unable to avail green finance from IDCOL.

Institutional Challenge

Questions related to institutional challenge were asked to government officials. The author's first question was whether they know about the term of green entrepreneurship.

Table 23. Knowledge about green/eco entrepreneurship

Opinion	No of Government officials	Percentage (%)
Yes	1	20%
No	4	80%
Total	5	100%

Source: Field Investigation.

This table illustrates that only 20% government officials said 'yes' when they were asked about green entrepreneurship. On the other hand, majority 80% green entrepreneurs answered 'No' when they were asked whether they knew about green entrepreneurship. It is surprising that most of the government officials who were interviewed have no knowledge about green entrepreneurship. The following table depicts their answer when they were asked whether they had received any motivation from government officials for promoting green entrepreneurship.

Table 24. Receiving motivation/instruction from government officials

Opinion	No of green entrepreneurs	Percentage (%)
Yes	0	0%
No	5	100%
Total	5	100%

Source: Field Investigation.

This table illustrates that none (0%) of the government officials said 'yes' when they were asked whether they motivate or instruct entrepreneurs to adopt green in their businesses. On the other hand, 100% green entrepreneurs answered 'No' when they were asked whether they motivate or instruct entrepreneurs to adapt green in their businesses. It is frustrating that most of the government officials do not give any motivation to entrepreneurs to adopt green in their business. During field work, the author asked green entrepreneurs about their opinion about the obstacles for the expansion of green entrepreneurship. One of the most important aspects of the survey was asking the question about constraints for the development of green entrepreneurship in Bangladesh's economic context. The data shows that the biggest obstacle for entrepreneurs is the financial obstacle to launch and run green business (50%) as well as the government's weaker policy towards the green entrepreneurial development (30%). Moreover, 20% respondents believe that entrepreneurship is still being perceived negatively by people even entrepreneurs themselves. The data shows that despite their interest in adopting innovations, entrepreneurs encounter a number of financial, institutional and behavioral challenges that always put them in trouble to cope with and overcome. When the author conducted detailed in-depth interviews, she obtained the following case stories. The author interviewed green entrepreneurs to understand the challenges of adopting green entrepreneurship in Bangladesh from their perspective.

Case Story 1: Became a Green Entrepreneur by Choice

Mr Tarikul islam (live at Basua area of Rajshahi city) said

> "I went to England to study Law. I had a chance to work part time in a company named Pindon energy. The company was German based and has a branch office in London. The company generated electricity through using waste which means they work on renewable energy. I wanted to earn through self-employment once I returned from England. I wanted to apply my green idea in Bangladesh that I learned from England. I wanted to work on renewable energy because I had some prior knowledges about renewable energy. I went to Banks and Infrastructural Development Company Limited (IDCOL) to avail finance. Although IDCOL is an investment company for green entrepreneurs, it does not provide financing to anyone living outside of Dhaka, capital city of Bangladesh. Therefore, I could not avail funding from IDCOL."

Source: The picture captured during field visit

Picture 1. Bio gas plant.

Mr. Tarikul also went to Bangladesh bank, Eastern bank and Prime Bank but he understood that bankers had either little or no knowledge about green financing. When he went to the Department of Environment (DOE), he had been informed that no license is required for such business, the explosion department had informed him that approval and license are required, which again need to be arranged from Dhaka city only.

Case Story 2: Became an Entrepreneur by Getting Inspiration from Others

In the beginning, Mr. Johirul thought that he would work with municipal waste for generating renewable energy. He went to the City Corporation but the Chief Conservancy Officer of the City Corporation informed him that they could provide a dumping place but they will not be able to invest for green business. Moreover, the problem remains in separating rotten and non-rotten municipal solid waste. The separating machine is very expensive that he could not afford to buy.

Source: Picture captured during field work.

Picture 2. CNG and Bio fertilizer plant.

Mr. Johirul futher added that

"people have the idea that biogas can only be produced from cow dung but the experiment shows that the biogas produced from cow dung is very small in amount only 16-22 cubic meter from a ton; anyway, biogas produced from municipal waste or other renewable waste products can produce more biogas. But he could not produce bio gas from municipal waste because of the lack of financial supports."

Case Story 3: Becoming Green Entrepreneur Is Extremely Challenging

Mr. Kamal stated that

"I wanted to become green entrepreneur for more than a decade. In the beginning, I went to Rajshahi sugar mill to collect waste to produce bio gas. Sugar mill produces three types of wastes such as bogus, molasses and press mud. Among these three types of wastes, press mud is best to produce bio gas. Then I started buying press mud from Rajshahi sugar mill. But I had to spend a lot of money for collecting press mud from sugar mill. Although Bangladesh sugar corporation does not use this waste, they charge 1050tk for per ton waste. Furthermore, I could not do the thermophilic digestion because of limited funding. I have been successfully producing CNG gas and bio fertilizer from throwaway waste products."

Source: Picture captured during collecting data.

Picture 3. Fertilizer produced from biogas plant.

Mr. Kamal contacted a successful businessman, Salam, for his advice who uses cow dung for producing bio gas. However, he does not know that bio gas can be produced from other materials as well. From this incidence, it is obvious that the green entrepreneurs do not know that bio gas can be produced from other raw materials. Most of the green entrepreneurs are completely ignorant about it. Mr. Kamal also informed that Nabil group who produced a lot of waste from corn. These corns can be used for producing bio gas. However, most of the big industries are not concerned about waste management. There is a big communication gap exists between those businessmen who are producing waste and the green entrepreneurs who want to use this waste to produce environment friendly goods. He stated that

> "one of the core challenges that I found that department of explosives and department of agriculture had expressed their inability to issue me permission letters to produce the gas and the bio fertilizer in a commercial basis. I wanted to take municipal waste from Rajshahi City Corporation to produce bio gas. But I could not get approval from Rajshahi City Corporation for using these municipal wastes. I failed to avail support from anywhere I started my work at my own place."

Case Story 4: Let There Be Light

Mr. Abdus Salam is an engineer by his profession. He said that

> "I bought a cow that cost 1.5 lakh Tk. in 1983. Then the number of cows increased to fifteen. However, the local people started complaining for the cow dung produced from a dairy firm. Then I started thinking about whether I can do something with this cow dung. Then I had started gaining knowledge regarding the production of biogas by using cow dung. Then I contacted RDA for necessary assistance. I build up a research lab before launching my green business. At some point, I bought a compressor machine from Chittagong. Then I started supplying biogas costs 50tk per Leguna[1]. But after three months, the department of explosives informed me that I should not have sold biogas this way

[1] Leguna is a local Vehicle

before getting approval. I achieved 23 lac 30 thousand taka from the government and I am hoping to achieve 50 thousand taka soon. And again, I am hoping to start my biogas plant soon by 2020".

He compared his business with others and came up with the following points

1. This is his own product, therefore he is earning more profit
2. Ten Kg. cow dung produces biogas worth 12.50 tk and can be sold at 80tk which is highly profitable
3. Other entrepreneurs are using stock gas but his is producing renewable energy.

Source: Field Study.

Picture 4. Machine which is used to produce biogas.

Mr. Abdus Salam talked about two other cases that adopted green in their business. In Thakurgaon, six hundred thousand Taka has been invested in producing biogas. But they are only producing biogas from municipal waste. But if some experienced people got their appointment in this project, then it would have been possible to supply gas with minimum cost. There are around 65,000 bio gas plants exists in India and 1 lac 50 thousand in China. He again said in Jamalpur,

> "Tauheedul islam has been recognized as one of the best young green entrepreneurs who has taken initiatives to produce oil through the plastic. He is quite successful in his green business. He has achieved government approval for his green project which helps him to sustain himself in the market. Therefore, it is very important to achieve government support for this type of green business."

Case Story 5: Not Adopting Green in Business

The owner of Nabil group started their work in 2005. In the beginning, they achieved 50,000tk loan from Islami Bank for their business. They took different initiatives for different businesses gradually. These are automatic rice mill, cold storage, flour mill etc. Their main business is corn. They collect corn from various places and produce many products by using corn. They produce various types of waste. From rice two types of wastes are being produced husk and burn. From wheat, two types of wastes mud and dust are being produced. He identified following challenges of green entrepreneurship:

Box 3. Challenges of Green entrepreneurship

- Very few people in Bangladesh want to be entrepreneurs let alone green entrepreneurs.
- Bangladeshi People hardly have risk taking capacity
- Lack of awareness
- Green entrepreneurs hardly get approval of their businesses
- Technological equipment that are required for promoting green entrepreneurship are very expensive
- Green entrepreneurs hardly receive any training

Source: Field Study.

On the other hand, the respondents shared some prospects of green entrepreneurship. They stated that biogas can be produced from potato waste but green entrepreneurs are not doing it due to their limited knowledge. One respondent said

> "In Mayapur Pilgrim, India 50,000 to 1,00,000 people have their lunches every day. They are producing biogas by using their food waste. They meet up their gas demand from biogas and they are also producing electricity by using their produced gas. He further added "an entrepreneur came from Saudi Arabia to launch bio plant in Padma Char land by using cow dung. He collected cow dung from local people for producing electricity and then he supplied electricity to local people. However, in char land, there is acute demand for cooking fuel. Therefore, Char people started using cow dung as cooking fuel. After some days, his bio plant project has been stopped. He supplied electricity to 272 houses in three months but he had to stop his project due to the limited supply of cow dung."

From the field study, the author was informed by most of the respondents that government officials should inform entrepreneurs that biogas can be produced through various types of wastes. Because people only know that biogas can be produced from cow dung. Therefore, more training is needed from the government side. An estimated 80 million tons of biomass fuel is being used annually for cooking, producing huge quantities of pollutants. About 30 million households in rural Bangladesh use biomass fuel for cooking. LPG and biogas are potential clean-burning fuels. Biogas generated from cattle/buffalo dung and poultry droppings is promising clean burning fuel. Considering the cattle/ poultry ownership, there is a potential of about 3 million small-sized biogas plants in the country; such plants could also be established in 33 dairy/poultry farms. Many organizations are currently involved in disseminating biogas technology. As of December 2012, a total of 65,317 biogas plants have been installed in the country; 26,311 by IDCOL (Hamid et al. 2013).

Case Story 6: A focus on Reduce, Reuse and Recycle

The author conducted an in-depth interview with the chief conservancy officer of Rajshahi City Corporation. He started his talk with three basic concepts of waste management such as reduce, reuse, and recycle. He said in the beginning, waste should have been reduced and left waste must have been reused and at last, waste will be recycled through which new products can be produced. However, many people are working with these wastages throughout the world, but people hardly get any recognized solution. Three 'R' can be the easy solution to get rid of wastage. At first, reusable waste should be stored, and excess waste needs to be recycled.

He said

> "in European countries, if anyone can provide five free packets of products, a new product will be delivered. They reuse these empty packets, and this could happen to our white detergent as well. Besides this, in South Korea, the 'sanitary land field' has become very successful. Since our country is very small, we cannot launch this. It requires a big space. The dumps are placed in various layers below the land and after a certain period, these dumps are turned into mud. But heavy infrastructure cannot be built upon this land."

In Rajshahi City Corporation, there is a waste collection system for managing waste but here the process of dumping is crude dumping as they do not have a recycling system. Besides, crude dumping is very common throughout the world. Waste is being kept in a specific place, the waste becomes rotten and turns to fertilizer. The people of Bangladesh are hardly become successful in producing renewable energy. If waste technology is successful, it will be successful too. But if this project becomes successful, another problem may arise that people will be interested in Polyethnic production which might be a big threat to the environment.

The chief conservancy officer declares that although the fertilizer is being produced from waste, there are many complaints against these fertilizers. But the fertilizer produced from waste cannot meet the standard criteria of fertilizer of Bangladesh. Therefore, they cannot get approval from the Government. He said

> "bioenergy can be produced from waste but it is not financially sustainable. Germany is the pioneer of energy production from waste. China tried these many times but none of the countries became successful. Because the process is very expensive but without government subsidy, it would never be successful."

He informed the author that recently a new contract has been signed between Rajshahi City Corporation and a Canadian citizen. He opined that 1300-liter dissel can be produced from one Ton Polithin and it would be economically sustainable. His American company's name is waste technology. This company has been successful in its pilot experimental project. If this project is implemented, this will be the first renewable plant that will be economically sustainable. This pilot project will be launched in the next two and a half years and it will work with 10-ton waste every day. If it becomes successful, the City Corporation will expand the program in a bigger context. Rajshahi City Corporation has signed a contract with waste technology.

According to this contract

1. RCC will provide space and waste
2. Company will use their technology and will invest
3. Profit will be shared once it is implemented

It is found from a field study that in the case of fertilizer selling, it cannot meet up the standard of Government policy regarding fertilizer. Therefore, these green entrepreneurs have restrictions on selling fertilizer as they cannot meet the standard set by the Government. Due to these restrictions, other organizations are not investing because no organization other than the government will invest without profit. In Rajshahi City Corporation, crude dumping takes place beside the city hut in Rajshahi. India has introduced biogas plants through waste a long ago but due to financial constraints, many of these gas plants have been abolished. If financial support is given, these biogas plants might be started again but most importantly people's mindset needs to be changed.

According to the chief conservancy officer of RCC,

> "Government is very much alert about green entrepreneurship. If it continues we will definitely notice some changes in the near future. In the proposed modern waste management structure, it will be seen how can these wastes be effectively used. He again said that rotten waste is needed

> for producing biogas. However, it is highly needed to separate the unusable waste. But entrepreneurs are not interested in this project because separating machine is very expensive."

The book argues that in Dhaka city, for example, the quantity of solid waste generated at present varies between 3000 to 3500 tons per day. Dhaka City Corporation is capable of collecting only 50% of this waste, leaving the remaining half unattended. A part of this waste either remains in the streets or on nearly open ground. Some of the waste flows to the open drains and blocks the normal drainage flow. Production of energy from waste is not economically viable because its establishment cost is quite high which can hardly be recovered without Government subsidy. In India, there is a running project on producing different products by using waste recycling but it also has a government subsidy. While the author was looking at the prospects, she met with officials of DOE and was informed that the Government has taken various steps to protect the environment. For example, i) the government has shifted Tanary from Hazari bag to Savar so that it reduces the effect on locality, ii) zero discharge plan, iii) ETB, Water cycle refine the process.

While conducting a depth interview with a government official, Nil Raton Sarkar informed

> "Government is much aware regarding green business. A monitoring team has been constructed. The monitoring team will act if any pollution is seen anywhere. ETB is a technology through which polluted water is refined and prepared to use again. In every Industrial area, a pond needs to be constructed compulsorily. Dirty water will be stored in the pond and then the dirty water is being refined through ETB to prepare it to reuse. No charge will be applied to those industries that use ETB. A project has been taken for reducing air pollution. As a part of this project, a flyover has been built. When the car goes slowly, an excess carbon comes out. That is why flyover has been constructed so that car can move speedily which can reduce pollution."

After an in-depth interview with the respondents, the author has found the following challenges and possibilities of adopting green entrepreneurship:

There is hardly any financial support from banks.

- Bank officials do not have adequate knowledge about green business. Even those who have knowledge about it are not willing to issue a loan in this sector.
- Big companies do not know who are interested in the field of waste management.
- Entrepreneurs do not know which waste can be used for producing biogas or renewable energy.
- There are some procedural challenges associated with green entrepreneurship. Therefore, entrepreneurs become frightened to come in green entrepreneurship.
- The entrepreneurs who had an interest in green business have become demotivated because of a lack of funding opportunities.
- Big companies produce huge amounts of waste but they have no knowledge and interest in using these wastes to produce renewable energy.
- The activities of SERDA (sustainable and renewable energy development authority) are only limited to seminars and workshops. There is no application of their activities.
- There is a big information gap between the Department of Environment and the forest and explosion department, which is a regulatory agency responsible for regulating commercial explosives in Bangladesh and is connected with the ministry of power, energy, and mineral resources.
- The entrepreneurs have limited knowledge about green business.
- The government takes insufficient initiatives to encourage entrepreneurs to launch and promote green businesses.

- Entrepreneurs' limited knowledge regarding green business is also a big challenge. People know that biogas can only be produced from cow dung but it can be produced from press mud, corn, etc.
- It is highly difficult for green entrepreneurs to get approval from the Sustainable and Renewable Energy Development Authority (SREDA)
- They are not getting approval for selling fertilizer that has been produced from biomass.
- It is extremely difficult for entrepreneurs to avail green finance.
- It is awfully difficult for entrepreneurs outside Dhaka to avail of loans from infrastructural Development Company Limited (IDCOL) or any other companies.
- It is also very challenging to get approval from the department of explosives unless entrepreneurs use the latest technology which is very difficult to avail of.
- Entrepreneurs are losing interest in the green business because it is very difficult to gain permission from the Government
- Another big challenge is a financial challenge. Without Government subsidy, entrepreneurs can hardly become successful.
- The green entrepreneurs are producing fertilizer as a by-product of biogas but they are unable to get Government approval. They can hardly meet the standard set by the Government.
- People's mental model needs to be changed because people may not be interested to use gas produced from waste

Installation of a biogas plant typically takes several weeks. For rapid dissemination, it is important to generate awareness among key stakeholders, especially women, regarding the availability/cost of technology, health advantages, and other benefits (e.g., quicker/easier cooking, reduction in workload for women). It is also important to take measures to ensure low-cost fodder and veterinary support for livestock in rural areas to encourage dairy at the household level.

POLICY SUGGESTIONS

As the author argued several times in the book that green entrepreneurship is one of the key methodologies of mitigating climate change. And both are extremely essential to achieve sustainable development. Environment-friendly products must be encouraged with the support of the Government subsidy. But the whole process should be controlled properly. With the contribution and support of both government and non-governmental organizations, environmental-friendly products should be encouraged. To increase the awareness of society regarding green entrepreneurship, social responsibility projects can be prepared together with role models. The companies that have a good reputation regarding green entrepreneurship from abroad and the companies that have an intention to perform green entrepreneurship from Bangladesh may come together and share their ideas. Owing to collaboration between universities and industry, environmental technology should be improved and also, green entrepreneurs must gain this technology at a low cost. Some specific policy suggestions are as follows:

a. Government should carefully monitor and supervise the green banking practices in Bangladesh.
b. Bangladesh bank must monitor the adherence of green banking guidelines by the commercial and non-commercial banks.
c. Government must encourage and aware of general people about green business and green banking through electronic and print media.
d. Coordination among concerned authorities is highly needed.
e. It is very essential to create awareness speedily and build capacity effectively.
f. Immediate concentration on sectoral lending policies and procedures are highly needed.
g. It is very essential to shift different categories of industry (such as garments, textiles, and tannery) to an appropriate location.

h. Awareness of top management is highly needed and it is indispensable to encourage borrowers to go green.
i. It is highly needed to apply green banking and implement environmental risk management (ERM) guidelines efficiently.
j. it is also required to develop a culture within the organization based on environmental governance.

CONCLUSION

Green entrepreneurship is the activity of consciously addressing environmental protection. It is something that has a net positive effect on the natural environment and at the same time is financially sustainable. Green entrepreneurship can provide solutions to youth unemployment and sustainability about the low entry-level requirements for entrepreneurs in several green sectors and their interest in innovative business solutions and sustainability deliberations. The book suggests that there are many challenges associated with green entrepreneurship. The core challenges include financial, institutional, and behavioral challenges. Lack of investors to invest in green businesses, bureaucratic hassles, delays in payments, and absence of application-based training platforms are some of the barriers to the development of green entrepreneurship in the Bangladeshi context, especially in the renewable energy sector as identified by the book. Hence, there is a need for government intervention as well as environmental awareness programs to develop a systematic approach towards green entrepreneurship. Engagement of a variety of stakeholders including various departments of the Government, NGOs, and utilities across Bangladesh would promote the development of renewable energy technologies. Then only we can expect to mitigate the adverse impacts of climate change in the long run. The book concludes that there is an immense opportunity for green entrepreneurs in the changing green market scenario in Bangladesh and the other way around, green entrepreneurs serve as drivers of change as they contribute to nurturing people's mindsets towards greener thinking and consumption, further boosting the dual effect of environmental and economic gains.

REFERENCES

Afrin S., Islam, N. and Ahmed, S. U. (2010). Micro Credit and Rural Women Entrepreneurship Development in Bangladesh: A Multivariate Model. *Journal of Business and Management.* 16, 1:14.

Adedeji, O. Reuben, O. and Olatoye (2014). Global Climate Change. *Journal of Geoscience and Environment Protection*. 2, 114-122.

Aktaruddin, M. (2000). Socio-Economic Profile of Industrial Entrepreneurs: A Study on North Region of Bangladesh. *A Journal of Banking and Finance*. 25, 39-53.

Aktaruddin, M. (1999). A Case study of some Successful and Unsuccessful Entrepreneurs in Bangladesh. *Journal of Business Administration.* 25(1&2), 89-126.

Akter, T. (2009). *Migration and living condition in urban slums; Implication for food security*, Dhaka: Unnayan Onneshan. retrieved on 3rd October ,2019 from .www.Unnayan.org/reports/migration. and.living.conditions.in.urban.slums.pdf

Allam, H. (2011) *Green Entrepreneurship, Catalyzing the transition to green economies*. Retrieved on 3rd September, 2020 from https://www.scribd.com/document/95044655/6-Green-

Alwakid, W., Aparicio, S. and Urbano, D. (2020). Cultural Antecedents of Green Entrepreneurship in Saudi Arabia: An Institutional Approach. *Sustainability*, 12, 1-20.

Aminuzzaman, M. S. (1991). *Introduction to social research*. Dhaka: Bangladesh Publishers.

American Psychological Association (2011). *Psychology and Global Climate. Addressing Multifaceted phenomenon and set of challenges*. Retrieved on 9th October, 2020 from http://www.apa.org/science/about/publications/climate-change.pdf.

Anabaraonye, B., Chukwuma, B. and Eriobu, C. M. (2019). Green Entrepreneurial Opportunities in Climate Change Adaptation and Mitigation for Sustainable Development in Nigeria Entrepreneurial Opportunities in Climate Change Adaptation and Mitigation for Sustainable Development in Nigeria. *Journal of Environmental Pollution Management* 2, 102.

Antonopoulos, R. (2009). *The unpaid care work - paid work connection* Working Paper No. 86, Retrieved on 1st December, 2020 from https://www.ilo.org/wcmsp5/groups/public/---dgreports/---integration/documents/publication/wcms_119142.pdf.

Asaduzzaman, M., Barnes, D. B. and Khandker, S. R. (2010). *Restoring Balance Bangladesh's Rural Energy Realities.* World Bank Working Paper No. 181 retrieved on 10th November from https://www.esmap.org/sites/esmap.org/files/P111450_Bangladesh_Restoring%20Balance-Bangladesh%27s%20Rural%20Energy%20Realities_Khandker.pdf.

Auerbach, F., & Silverstein, L. B. (2003). *Qualitative data: Introduction to coding and analysis.* New York: New York University Press.

Banerjee, S., & Dutta, A., B. (2017). Green HRM: A brief review. *Rainbow Multidisciplinary Pre-Reviewed Annual Journal*, 3(1), 1-5.

Bazeley, P. (2007). *Qualitative data analysis with NVIVO*. London: Sage Publications.

Begum, A. (1993). *Government-NGO interface in development management: Experiences of selected collaboration models in Bangladesh.* Dhaka: A. H. Development Publishing House.

Beg, N., Morlot, C. J. Davison, O. Okesse, Y. A. Tyani,L. Denton, F. (2011). *Linkages between climate change and sustainable development.* Retrieved on 10th April, 2020 from https://www.tandfonline.com/doi/abs/10.3763/cpol.2002.0216.

Berg, B. L. (2009). *Qualitative Research Methods for the social sciences.* California State University, USA.

Bhattacharya, A. Oppenheim, J. and Stern, N. (2015). Driving Sustainable Development Through better infrastructure: Key elements of a transformation program. *Global Economy and Development working paper* 91. 11-38.

Bizikova, Robinson, J. and Cohen, S. (2007) Linking climate change and sustainable development at the local level, *Climate Policy*, 7:4, 271-277 To link to this article: https://doi.org/10.1080/14693062.2007.9685655.

Brundtland, G. H. (1987). *Our Common Future: Report of the World Commission on Environment and Development retrieved on 1st November,* 2020 from http://www.un-documents.net/our-common-future.pdf.

Bryman, A. and Bell, E. (2003). *Business Research Methods*, Oxford: Oxford University Press.

Burzyńska, D., Jabłońska, M., and Dziuba, R. (2018). Opportunities and conditions for the development of green entrepreneurship in the polish textile sector. *Fibres & Textiles in Eastern Europe*, 26(2), 13–19.

Čekanavičius*, L, Bazytė, R and Dičmonaitė, A (2014). Green Business: Challenges and Practices, *ECONOMICA*, 93 (1).

Center for Climate change and health (2015). *Public Health Institute's Center for Climate Change & Health is dedicated to tackling the challenge of climate change while creating more vibrant, healthy and equitable communities throughout the US and around the world.* Retrieved on 1st December, 2020 from https://climatehealthconnect.org.

Creswell, J. W. (2012). *Educational research: Planning, conducting and evaluating quantitative and qualitative research.* Upper Boston, MA: Pearson Education.

Dankelman, I. (2010). Climate change: Learning from gender analysis and women's experiences of organising for sustainable development. *Gender and Development*. 10 (2), 21-29.

Davies, A. R. and Mullin, S. J (2011). *Greening the economy: Interrogating sustainability innovations, beyond the mainstream*. 11(5), 739-81.

Denzin, N. K., and Lincoln, Y. S. (2000). *Handbook of qualitative research.* California: Sage Publishing.

DFID (2011). *Bangladesh Operational Plan.* Retrieved on 5th January, 2020 from www.dfid.gov.uk/Documents/publications1/op/Bangladesh-2011.pdf.

Dion, J. (2012). *Developing Financeable NAMAs A Practitioner's Guide*, International institute for sustainable development report Retrieved on 22th September from https://www.nab.vu/sites/default/files/nab/documents/06/05/2013%20-%2010:35/developing_financeable_namas1.pdf.

Dion, J., Murphy, D., Harris, M. and Sawyer (2012). *Low-Carbon, Climate-Resilient Development NAMA Concepts for Bangladesh,* Retrieved on 3rd September, 2019 from http://oldweb.lged.gov.bd/UploadedDocument/UnitPublication/1/914/NAMA%20Concepts%20for%20Bangladesh.pdf.

Dubash, N. K. (2019). *Revisiting climate ambition: The case for prioritizing current action over future intent.* Retrieved on 1st February, 2020 from https://onlinelibrary.wiley.com/doi/full/10.1002/wcc.622.

Elliot, J. A. (2013). *An Introduction to sustainable development*. Routledge, London and New York.

Falle, S., Rauter, R. and Baumgartner (2018). Sustainability Management with the Sustainability Balanced Scorecard in SMEs: *Findings from an Austrian Case Study*, Sustainability, 8(6), 545.

Farinelli and Bottini, M. (2011) *Green entrepreneurship: A missing link towards a greener economy. Positive externalities of green entrepreneurship and innovation.* working paper number 2013/41, retrieved on 4th March, 2019 from https://www.researchgate.net/

publication/285170513_Green_entrepreneurship_The_missing_link_towards_a_greener_economy.

Gast, J, Gundolf, K. and Cesinger, B. (2017). Doing business in a green way: A systematic review of the ecological sustainability entrepreneurship literature and future research directions, *Journal of Cleaner Production*. 147, 44-56.

Gevrenova T. (2015). Nature and characteristics of green entrepreneurship. *Trakia Journal of Sciences* 13 (2).

Green Climate Fund (2012). *Infrastructure Development Company Limited.* Retrieved on 14th November, 2019 from https://www.greenclimate.fund/ae/idcol.

Groom, A. J. R. (2001). *Getting to Go. The birth of United Nations system.* In Taylor, P. and Groom, A. J. R and (Eds). The United Nations at the millennium, CONTINUUM, London.

Gunawan and Fraser (2017). Exploring young and green entrepreneurship in Indonesia: An introduction. *Journal of Business Strategy*, 6(9),185-194.

Gupta, H. and Barua, M. K. (2018). A framework to overcome barriers to green innovation in SMEs using BWM and Fuzzy TOPSIS Himanshu. *Science of the total environment*, 633,122-139.

Habibullah, M. (1987). *Entrepreneurship, Training and Development*, Bureau of Business Research, 47-58.

Halder, S. (2018). Green entrepreneurship in the renewable energy sector: A case study of Gujrat, *Journal of science Technology and policy limited,* 22-15.

Hallegatte, S. and Rozenberg, J. (2017). Climate change through a poverty lens, *Nature Climate Change*, 7, 250-256.

Hancock, B. (1998). *Trend focus for research and development in primary health care: An Introduction to qualitative research.* Retrieved on 14th July, 2020. from http://faculty.cbu.ca/pmacintyre/course_pages/MBA603/MBA603_files/IntroQualitativeResearch.pdf.

Hansen and Schaltegger (2018). Sustainability Balanced Scorecards and their Architectures: Irrelevant or Misunderstood? *Journal of Business Ethics,* 150(4), 937–952.

Helal, M. and Hossain, M. A. (2013). Four Decades of economic development of Bangladesh: An assessment. *Journal of the Asiatic Society of Bangladesh*. 58(2), 335-362.

Hensher, D. A. (2008). Climate change, enhanced greenhouse gas emissions and passenger transport – What can we do to make a difference? *Transportation Research Part D: Transport and Environment.* 13(2), 95-111.

Hopwood, B. Mellor, M. and O'Brien (2005). Sustainable development: mapping different approaches. *Sustainable Development*, 13 (1), https://doi.org/10.1002/sd.244.

Holtzhausen, S. (2001). Triangulation as a powerful tool to strengthen the qualitative research design: The resource-based learning career preparation program (RBCLCPP) as a case study. *Paper presented to the Higher Education Close Up Conference* 2. Lancaster University, Lancaster, UK.

Hossain, K. S. and M. Rahman (1999). Role of Grameen Bank in Entrepreneurship Development: A Study on Some Selected Entrepreneurs. *Islamic University Studies* (Part C), 2, 7-13.

Huq, S. (2001). *Climate Change and Bangladesh*, Science, 294 (5547):1617, DOI: 10.1126/science.294.5547.1617.

IPPC (2014). *Climate Change 2014 Mitigation of Climate Change Summary for Policymakers Technical Summary Part of the Working Group III Contribution to the Fifth Assessment Report of the Intergovernmental Panel on Climate Change*, retrieved on 9th December, 2019 from https://www.ipcc.ch/site/assets/uploads/2018/03/WGIIIAR5_SPM_TS_Volume-3.pdf.

IPCC (2007). *Climate Change 2007: The Physical Science Basis. Contribution of Working Group I to the Fourth Assessment Report of the Intergovernmental Panel on Climate Change (IPCC)* [Solomon, S., D. Qin, M. Manning, Z. Chen, M. Marquis, K. B. Averyt, M. Tignor and H. L. Miller (eds.)]. Cambridge University Press, 996 pp.

Islam, A. and Islam and Aktaruzzaman (2011). Effect of Entrepreneur and firm characteristics on the business success of small and medium enterprises (SMEs) in Bangladesh, *International Journal of biometrics*. 6, 289

Islam, M. N. (2008). *An introduction to research methods: a handbook for business and health research*. Dhaka: Mullick & Brothers.

IUCN (2020). *Gender and the environment: What are the barriers to gender equality in sustainable ecosystem management?* Retrieved on 22nd. July from https://www.iucn.org/news/gender/202001/gender-and-environment-what-are-barriers-gender-equality-sustainable-ecosystem-management.

Jackson, T. (2009). *Prosperity without growth? The transition to a sustainable economy.* Retrieved on 3rd November, 2019 from http://www.sd-commission.org.uk/data/files/publications/prosperity_without_growth_report.pdf.

Jonsson, S. A. (2011). Virtue and vulnerability: Discourses on women, gender and climate change. *Global Environmental Change*. 21 (2), 744-751.

Khanka, S. S. (2002). *Entrepreneurial Development*. New Delhi, India: S. Chand Group.

Klarin (2018). The Concept of Sustainable Development: From its Beginning to the Contemporary Issues, *Zagreb International Review of Economics & Business*, 21, 67-94.

Koe, W. L, Omar, R. and Saari, J. R. (2015), Factors Influencing Propensity to Sustainable Entrepreneurship of SMEs in Malaysia. *Procedia - Social and Behavioral Sciences*, 172, 570-577.

Kucukoglu, M. T. and Pinar, R. I. (2015). Positive Influences of Green Innovation on Company Performance, *Procedia - Social and Behavioral Sciences*, 195, 1232-1237.

Kusat, N. (2013) Green Economy for Green Sustainability: Advantages And Disadvantages- Turkey Review, *Journal Of Yasar University*. 29 (8), 4896-4916.

Lele, S. (2001). - *Sustainable Development: A Critical Review, World Development,* 19, 607-621.

Mathur, S., and Tandon, N. (2016, December). Green entrepreneurship: the emerging paradigm for sustainable growth and development in India-A

Study of the Millennials. *Indian Journal of Science and Technology,* 45, 25-30.

Mensah, J. (2019) Sustainable development: Meaning, history, principles, pillars, and implications for human action: Literature review, *Cogent Social Sciences,* 5 (1), 1-21.

Merriam, S. B. (1988). *Qualitative research: A guide to design and implementation*. San Francisco: Jossy- Bass, San Francisco.

Miles, M. B., & Huberman, A. M. (1994). *Qualitative data analysis: An expanded source book.* London: Sage Publications.

Mishra, P and Sharma, P. (2010). Green Marketing in India: Emerging Opportunities and Challenges. *Journal of Engineering, Science and Management Education.* 3, 9-14.

Ministry of Environment and Forests (MoEF) (2005). *National Adaptation Program of Action,* 2009 (Updated Version). Dhaka: Government of the People's Republic of Bangladesh.

Mou, I. and Azeez, A. (2019). Green Entrepreneurship: Literature Review and Agenda for future research. *International Journal of Entrepreneurial Knowledge*. 2, 17-29.

Mrkajic, B, Murtinu, S. and Scalera, V. G. (2017). Is green the new gold? Venture capital and green entrepreneurship, *Small business economics,* doi 10.1007/s11187-017-9943-x.

Muff, K. Kapalka, A., and Dyllick, T. (2017). The Gap Frame—Translating the SDGs into relevant national grand challenges for strategic business opportunities. *International Journal Management Education.* 15, 363–383.

Nachmias, F. C., & Nachmias, D. (1996). *Research methods in the social sciences*. London: Arnold.

Nature (2013). *Sustainable development goals for people and planet,* Retrieved on 12th September 2020 from https://www.nature.com/articles/495305a.pdf.

Nawaz, F. (2015). Violence, Gender and Environmental Justice in India: A Critical Analysis, *Scholars Bulletin.* 1(4), 96-101.

Nawaz, F. (2011). Problems of Woman Entrepreneurship Development in Bangladesh: A Case Study of RAKUB, *Pertanika J. Soc. Sci. & Hum.* 20 (3), 601 – 614.

Nawaz, M. S. (2007). *An evaluation of microcredit as a strategy to reduce poverty and empower women: A case study of three microcredit programs in Bangladesh*. Unpublished doctoral thesis, Flinders University - Adelaide, Australia.

Nikolaou, I.E, Tasopoulou, K. and Tsagarakis,K (2018), A Typology of Green Entrepreneurs Based on Institutional and Resource-based Views, *The Journal of Entrepreneurship,* 27(1), 111–132.

OECD (2011). *Towards green growth A summary for policy makers retrieved on 4th March,* 2020 from https://www.oecd.org/greengrowth/48012345.pdf.

O' Neil K. and Gibbs, D (2016). Rethinking green entrepreneurship - Fluid narratives of the green economy. *Environment and planning* A. 48 (9). 10.1177/0308518X16650453.

Pouspourika, K. (2020). Green Entrepreneurship. Sustainable development for business. *Entrepreneurship*.

Rahman, M., Ibrahim, Hossain; and Miah, A. S. (2000). Problems of Women Entrepreneurship Development: A Study of Grameen Bank Finance on Some Selected Areas. *Islamic University Studies* (Part-C), 3, 124-128.

Rahman, A (1999) *Women and Microcredit in Rural Bangladesh -- An Anthropological Study of Grameen Bank Lending*. Boulder, CO: Westview Press.

Rudra, S., Kurian, O.C. (2018). Progress Tracking of Health-Related SDGs: Challenges and Opportunities for India. *Asian Journal of Public Affairs*. 10, 24–52.

Sachs, P. J. (2012). *From Millennium Development Goals to Sustainable Development Goals*. Retrieved on 3rd September 2019 from https://www.thelancet.com/journals/lancet/article/PIIS0140-6736(12)60685-0/fulltext.

Saleh, A. (1995). A Profile of the Women Entrepreneurship in Bangladesh," *Journal of Business Studies*. 16 (1), 159-170.

Sambrani, V. N. G*overnment Policies and Rural Entrepreneurship – Case Study of a Horticulture Entrepreneur* retrieved on 3rd April, 2020 from http://www.publishingindia.com/GetBrochure.aspx?query=UERGQnJvY2h1cmVzfC8zNDI5LnBkZnwvMzQyOS5wZGY.

Sen, A. (2004) *The State of World Population 2001*, Australian Reproductive Health Alliance, Weston.

Sharma and Kushwaha (2015) Green initiatives: a step towards sustainable development and firm's performance in the automobile industry, *Journal of Cleaner Production*, 3, 1-4.

Shahidullah, A. K. M & Haque, C. E. (2015), Green microfinance strategy for entrepreneurial transformation: validating a pattern towards sustainability, *Enterprise Development and Microfinance*. 26 (4), 325-342.

Shi, L. Han, L. Yang, F. and Gao, L. (2019). The Evolution of Sustainable Development Theory: Types, Goals, and Research Prospects. *Sustainability,* 11(24), 7158.

Silajdzi, I, Kurtagi, S M and Vucijak, B, (2015), Green entrepreneurship in transition economies: a case study of Bosnia and Herzegovina, *Journal of Cleaner Production*. 88, 376-384.

Simanowitz, A. (2003). Microfinance poverty and social performance: Overview. *IDS Bulletin*. 34(4), 1-5.

Smith, A. and Grin, J. (2010). *Innovation studies and sustainability transitions: The allure of the multilevel perspective and its challenges*. 39, 435-448.

Soubbotina, T. P. (2004). *Beyond Economic Growth: An Introduction to Sustainable Development.* The World Bank Washington, D.C. USA.

Sustainable Development Solution Network (2015). *Sustainable Development Report 2020.* Retrieved on 3rd November, 2019 from https://www.unsdsn.org.

Swathi, G. S. (2017). Women Empowerment in Scheduled Caste Women: Problems and Challenges. In Rao, T. V. (Eds) *Sustainable development. In deal International E-Publication*. India.

Swart, R. Robinson, J. and Cohen, S. (2003). Climate change and sustainable development: expanding the options. *Climate Policy*. 3(1), 19-40

Thang, N. N, Quang, T. and Son, N. H. (2013), Knowledge creation and green entrepreneurship, *Asian Academy of Management Journal*, Vol. 18(2):129-181.

Tharindu, P. K and Koggalage, R. (2020). Dynamics of Green Entrepreneurship Concept: A Review of Literature. *Journal of Multidisciplinary Engineering Science and Technology (JMEST)*. 7 (8), 12529-12533.

The Asia Foundation (2012). *Building climate resilience in Bangladesh*. Retrieved on 22nd November, 2019 from https://asiafoundation.org/ resources/pdfs/BGClimateChange.pdf.

The Declaration of Stockholm (1972), retrieved on 3rd January from https://www.google.com/url?sa=t&rct=j&q=&esrc=s&source=web&cd =&ved=2ahUKEwjVsp-p5aLsAhWgH7cAHay7DrcQFjACegQIAhA C&url=https%3A%2F%2Fwww.ipcc.ch%2Fapps%2Fnjlite%2Fsrex% 2Fnjlite_download.php%3Fid%3D6471&usg=AOvVaw07cRcaLVdAt 45A7exTjasV.

UNEP (2011). *Towards a Green Economy: Pathways to Sustainable Development and Poverty Eradication.* Retrieved on 3rd September, 2020 from https://sustainabledevelopment.un.org/index.php? page=view&type=400&nr=7&menu=35.

United Nations Climate change (2019). *UN climate change conference.* Retrieved on 2nd October, 2020 from https://unfccc.int/cop25.

United Nations (2015). *List of SDG goals, targets and full text indicators covered in the proposed core basket of SDH action indicators.* Retrieved on 1st November 2020 from https://www.who.int/ social_determinants/CommissionSDG-goals-targets-indicators-app7.pdf.

Uslu, Y. D, Hancioglu, Y & Demir, E (2015) Applicability to Green Entrepreneurship in Turkey: *A Situation Analysis, Social and Behavioral Sciences*, Vol. 195: 1238 – 1245.

Voumik, L. C and Shah, M. G. H. (2014). A green Economy in the Context of Sustainable Development and Poverty Eradication: What are the Implications for Bangladesh? *Journal of Economics and Sustainable Development*. 5 (3), 119-131.

Warrick, Barrow and Wigley (1993). *Climate and sea level change: Observations, projections and implication*. Cambridge university press, Cambridge.

World Bank (2018). *Global Waste to Grow by 70% by 2050 unless urgent action is taken: World Bank report*. Retrieved on 11th November, 2019 from https://www.worldbank.org/en/news/press-release/2018/09/20/global-waste-to-grow-by-70-percent-by-2050-unless-urgent-action-is-taken-world-bank-report.

World Bank (2013). *World Development Indicators 2002*, retrieved on 23rd December, 2019 from https://elibrary.worldbank.org/doi/abs/10.1596/0-8213-5088-9.

Yin, R. K. (1994). *Case study research: Design and methods*. Thousand Oaks, CA: Sage.

Zahedi, A. and Otterpohl, R. (2015). *Towards Sustainable Development by Creation of Green Social Entrepreneur's Communities*, Procedia CIRP, 26, 196– 201.

Ziring, L; Riggs, Robert E.; and Plano, Jack C., "The United Nations: International Organization and World Politics" (2005). *All Books and Monographs by WMU Authors.* 308. https://scholarworks.wmich.edu/books/308.

ABOUT THE AUTHOR

Dr. Faraha Nawaz is an Associate Professor of the department of Public Administration, University of Rajshahi, Bangladesh. Dr. Nawaz has proved her excellence both in her basic and applied researches as well as in publishing. As an academic, she has devoted herself in undertaking wide-ranging research in social Science and has published more than 30 articles in peer reviewed SCOPUS indexed international journals, chapters in edited books, and three sole authored books published from Germany, UK and USA. Dr Nawaz is an active member of a number of international professional groups and networks such as NAPSIPAG, ISTR, EROPA, NZSAC, Dev Net and AWID, and has presented scholarly papers in various prestigious international conferences. She is a recipient of esteemed awards including gold medals, Charles Wallace award, Dev net Research award from New Zealand, Malaysian scholarly award, Australian Government award, etc. She has been awarded as best social science researcher 2020 from new science inventions. She was an international fellow of Watson Institute for International and Public Affairs, Brown University, USA and a visiting research scholar of Contemporary South Asian Studies program, School of Interdisciplinary area studies, University of OXFORD, UK.

Publications in Last 3 Years

Book Publication:

Nawaz, F. (2020) *Climate Change Mitigation and Green Entrepreneurship: Risks and Response*, NOVA science publisher, New York, USA. (To be published in December).

Nawaz, F. (2019) *Microfinance and Women's empowerment in Bangladesh: Unpacking the Untold Narratives*, Palgrave Macmillan Publishing, London, UK.

Chapters in Edited Books

Nawaz, F. (2020). Empowering marginalized women through entrepreneurship: Breaking the barrier, In Lokuge (Eds) *'Rural Entrepreneurship and Innovation in the Digital Era,'* IGI Global, Pennsylvania, USA

Nawaz, F. (2018). Conscientization among rural women: A case of Bangladesh, In Nilan (Eds) *'Conscientization,'* Routledge, Taylor and Francis Group, New York

Articles Published in Peer Reviewed Academic Journals

Nawaz, F. (2020) *Locked Up Under Lockdown the COVID-19 Pandemic and the Migrant Population*. Available at SSRN: https://ssrn.com/abstract=3631707 or http://dx.doi.org/10.2139/ssrn.3631707 (Dr AKM Ahsan Ullah and Diotima Chattoraj)

Nawaz, F. (2020). The Impact of Non-Go the Impact of Non-Government Organizations on Women's Mobility in Public Life: An Empirical Study in Rural Bangladesh, *Journal of International Women's Studies*. Vol 21(2):91-113.

Nawaz, F. (2020). Surrogacy led migration: Reflections on Policy Dilemmas. *Public Administration and Policy: An Asia-Pacific Journal*, Vol.23:1-15 (With Dr AKM Ahsan Ullah).

Nawaz, F. (2020) *Social research methods: migration in perspective, Migration Letters,* 17(2):351-362. (With Dr. AKM Ahsan Ullah, Md Akram Hossain and Mohammad Azizuddin).

Nawaz, F. (2020) Xenophobia in the GCC countries: migrants' desire and distress. *Global Affairs*, Vol 6(2):1-22 (With Dr. A.K.M. Ahsan Ullah, Dr. Shirley Chin Wei Lee, Dr. Noor Hasharina Hassan).

Nawaz, F. (2020) Challenges of public policy implementation: Consumers Rights Protection Act in Bangladesh. *Journal of Humanities and Social Sciences Research,* 2 (1): 51 – 60 (Dr. Nayan Kanwal).

Nawaz, F. (2019). Reintegration Challenges of Migrants in Bangladesh: A Study on Forced Returnee Women Migrants from Saudi Arabia. *Journal of Humanities and Social Sciences*, Vol 1(1):49-58 (with Tania Afrin Tonny)

Nawaz, F. (2019). Adverse Political Incentives and Obstinate Behavioural Norms, A Study of Social Safety Nets in Bangladesh. *International Review of Administrative Sciences,* Vol 26(1):70-91 (Dr. Abu Elias Sarker)

Nawaz, F. (2019). Pathways to Women's Empowerment in Bangladesh: Employment and Microfinance as Interventions, *Journal of Asian Education and Development Studies* (with Dr. Habib Zafarullah) https://doi.org/10.1108/AEDS-11-2018-0168

Nawaz, F. (2019). Clientelism, Patriarchy and Democratic Backsliding: Role of local Government elections in Bangladesh. *South Asian Survey*. Vol. 26 (1):70-91(Dr. Abu Elias Sarker)

Nawaz, F. (2018). Social Capital, Collective action and Women's empowerment, *South Asian Journal of Policy and Governance*, Vol. 42 (1).

Nawaz, F. (2018). An Analysis of the role of Community Empowerment Program in Combatting Domestic Violence against women: A Case Study of Bangladesh, *Pertanika Journal of Social Science and Humanities*. Vol. 26 (4): 2841 – 2856.

INDEX

B

C

E

G

M

N

R

S

W